The Institute of Biology's
Studies in Biology no. 96

Plant Growth Analysis

Roderick Hunt
B.Sc., Ph.D., M.I.Biol.

Independent Research Worker, Natural Environment Research
Council Unit of Comparative Plant Ecology, Honorary Lecturer
in Botany, University of Sheffield

Edward Arnold

© Roderick Hunt 1978

First published 1978
by Edward Arnold (Publishers) Limited,
41 Bedford Square, London WC1B 3DP

Boards edition ISBN: 0 7131 2695 7
Paper edition ISBN: 0 7131 2696 5

Printed and bound in Great Britain at
The Camelot Press Ltd, Southampton

General Preface to the Series

It is no longer possible for one textbook to cover the whole field of Biology and to remain sufficiently up to date. At the same time teachers and students at school, college or university need to keep abreast of recent trends and know where the most significant developments are taking place.

To meet the need for this progressive approach the Institute of Biology has for some years sponsored this series of booklets dealing with subjects specially selected by a panel of editors. The enthusiastic acceptance of the series by teachers and students at school, college and university shows the usefulness of the books in providing a clear and up-to-date coverage of topics, particularly in areas of research and changing views.

Among features of the series are the attention given to methods, the inclusion of a selected list of books for further reading and, wherever possible, suggestions for practical work.

Readers' comments will be welcomed by the author or the Education Officer of the Institute.

1978   The Institute of Biology,
                                        41 Queens Gate,
                                        London, SW7 5HU

Preface

In the very first volume of this series John Phillipson wrote 'The biological capacity of the earth depends ultimately on the energy received from the sun, and man, to satisfy amongst other things his demand for food, depends on the use to which this energy is put by living organisms.' The present booklet gives an introduction to the methods which are available for the quantitative analysis of a fundamental stage in this flow of energy: the growth of the whole, autotrophic plant in relation to its environment. In crop science, and in the study of natural vegetation, important information may be gained from a straightforward treatment of the simplest raw data on plant growth. Yet more elaborate and informative calculations and a more labour-saving approach to the easy ones are both within reach of many of the programmable calculators with which even quite modest laboratories are now equipped.

Dr Gillian Thorne kindly provided comments on an early draft. Additional help was given by Professor A. J. Willis, Dr H. Gretton and Dr A. J. M. Baker. The permissions of the authors and publishers of the various ilustrative examples are also gratefully acknowledged. Much of the booklet is based on lecture material from courses given at the Universities of Bristol and Sheffield.

Sheffield, 1978   R. H.

Contents

1 Growth and Growth Analysis

1.1 Growth as a unifying concept in biology

All living organisms are, at various stages in their life history, capable of change in size, change in form and change in number, given suitable conditions. These three processes together form an important part of the phenomenon of life itself and among natural systems help to distinguish the living from the non-living. The three are strongly interlinked and the term 'growth' may be applied to any or all of them. For these reasons a precise definition of what is meant by 'growth' is not at all easy. Definitions may range from an unequivocal statement about change in a specified dimension to a highly abstract state of affairs in which the verb 'to grow' means nothing more than 'to live' or even 'to exist'. No firm definition will be advanced to cover the use of the term in this booklet, other than to say that it will be used mainly to describe irreversible changes in size (however measured), often to describe changes in form and occasionally to describe changes in number, all as applied to the quantitative study of plant performance.

1.2 Growth in different organisms

This booklet is about analytical concepts and techniques that are designed to be applicable to a wide range of organisms. These have been developed so as to negate, as far as possible, the inherent differences in scale between contrasting organisms so that their performances may be compared on an equable basis. In Table 1 the rates of dry weight increase (irreversible growth in size) are given for a range of organisms grown under favourable conditions. Although much variation exists within these broad groups with respect both to the maximum rate of increase and to the environmental conditions in which this may be attained, one general conclusion is clear: the larger and more complex the living organism, the lower is the rate of dry weight increase possible, when expressed on a percentage basis. This trend is generally held to be due to the increased morphological and anatomical differentiation which is necessary to sustain life in large systems. This differentiation leads to translocatory pathways of increased length between the point of entry of raw materials into the organism and the site of its nucleoprotein replication.

Table 1   Rates of dry weight increase for a range of organisms grown under favourable conditions. (Developed from WILLIAMS, 1975.)

Organism	Percentage dry weight increase per day	Sources
Bacteriophage		1 ELLIS and DELBRUCK, 1939
Anti-*Escherichia coli* phage (1)	20 400	2 BROCK, 1967
Bacterium		3 OTSUKI, SHIMOMURA and TAKEBE,
Escherichia coli (2)	4 750	1972
Virus		4 WHALEY, 1961
Tobacco mosaic virus (3)	2 210	5 TRINCI, 1969
Yeast		6 GRIME and HUNT, 1975
Willia anomala (4)	1 400	7 RAJAN, BETTERIDGE and BLACKMAN,
Fungus		1973
Aspergillus nidulans (5)	860	8 COOMBE, 1960
Algae		
Chlorella (T × 7115) (4)	620	
Anabena cylindrica (4)	74	
Angiosperms – herbaceous		
Poa annua (6)	38.6	
Helianthus annuus (7)	29.0	
Nardus stricta (6)	10.1	
Angiosperms – woody seedlings		
Fraxinus excelsior (6)	12.8	
Trema guineensis (8)	5.31	
Acer pseudoplatanus (5)	4.85	
Gymnosperms – seedlings		
Picea abies (6)	6.00	
Pinus sylvestris (6)	5.13	
Picea sitchensis (6)	3.14	

The point that this table illustrates is that although the differences in organization among groups could scarcely be greater, calculations made in this way allow fair quantitative comparisons to be drawn.

1.3   An introduction to growth in higher plants

In a series of classic experiments performed at Poppelsdorf, West Germany in the 1870s, U. Kreusler and co-workers demonstrated that the growth of an annual plant under natural conditions followed a course that has since been recognized as typical of many. In Fig. 1–1a their data are given for the increase with time in mean dry weight per plant in *Zea mays* (maize) cv. 'Badischer Früh' grown in 1878 (KREUSLER, PREHN and HORNBERGER, 1879).

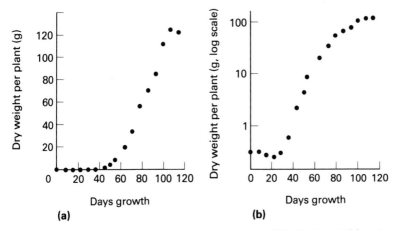

Fig. 1-1 Changes in dry weight with time in plants of 'Badischer Früh' maize grown at Poppelsdorf in 1878. (Plotted from data given by KREUSLER, PREHN and HORNBERGER, 1879.)

Because the changes in dry weight over the whole period are of the order of 370-fold, very little of the first phases of development is revealed in this simple plot of dry weight against time on an arithmetic scale. If these data are transformed to (natural) logarithms we can see more clearly what is happening (Fig. 1-1b). There is no special reason why natural, rather than common, logarithms should be used for this purpose, but since these data will be referred to again in a context that will require such a transformation then it is convenient to introduce it here.

We see that the plant shows no change in dry weight for the first 10 days or so. Then it actually loses weight until about 20 days have passed. Here is an example of a difficulty encountered in defining growth: no increase in weight has occurred but there has been considerable differentiation of leaf tissue in the young seedling (at the expense of total dry weight). From about 20 days the newly-differentiated leaves are able to contribute substantially to carbon assimilation and a so-called grand period of growth begins in which the unfolding of new leaves and an increase in total dry weight occur continuously. The plants flower at about 65 days and an increasing proportion of assimilate is now directed into the developing ear or cob with a corresponding tendency for the lower leaves to atrophy. Finally, at about 105 days, net dry weight increase ceases although growth in the sense of a continuing partition of dry weight into ears continues.

This pattern of growth, with great variation in the magnitude of the dry weight values, in the symmetry of the curve and in the time scale which it occupies, is general among annual plants grown in a productive environment. In perennial plants the pattern is similar at first but later, at

least in a temperate climate, dry weight increase proceeds in a series of
annual steps which may be linked by periods of negative growth in-
between. Naturally, the environmental conditions affect the magnitude of
growth at all stages.

1.4   The aims of this volume

Over the last sixty years a body of quantitative techniques has been
built up which allows the experimenter to derive important comparative
information about the growth of whole plants under natural, semi-
natural or artificial conditions. These techniques require only the
simplest of basic data, such as those described above, and have
collectively become known by the informal title 'plant growth analysis'.
This booklet is mainly concerned with ways of carrying out such analyses
– a kind of technical manual.

As in other scientific fields, these techniques have to be kept in
perspective and any suggestion of a technical cult avoided. To introduce
the first of several mechanical analogies; engineering problems are solved
not solely by the meticulously correct use of tools; a wider appreciation of
objectives is required. But, conversely, the project will suffer if the
engineer is so fixated by his objectives that he uses pliers to undo nuts and
a hammer to drive in screws. A more balanced approach is required, with
a proper alliance between purpose and method.

This volume is intended to provide a short, first acount of the technical
side of this balance upon which students, and others new to the field, may
build if required. The emphasis throughout is upon the use of the
concepts of plant growth analysis as comparative tools.

2 Collection of Basic Data

2.1 Design of experiments

2.1.1 General considerations

It is assumed that experimenters grow plants to test an hypothesis. Possibly we wish to show that one particular environment or management practice is or is not more suitable for a particular plant than another. Or we may wish to compare the performances of different species or varieties grown under the same conditions. Or we may just wish to explore and quantify the growth of a new experimental subject. In any of these cases a programme of activity can probably be designed that will give us the information that we are looking for through the medium of growth analysis.

Although the design of experiments proper is outside the scope of this booklet, the problems involved can be set down simply: Can we be sure of selecting the right experimental material? Can we use it in such a way that any differences in growth genuinely reflect the treatments applied? Can we sample the material in such a way as to get down on to paper an accurate version of these differences? Can we handle these data in a way that illuminates our understanding of the real events that have occurred and provides the basis for the acceptance or rejection of our hypothesis? An introduction to the important processes of hypothesis-generating, variability in material, sampling of populations, replication, randomization and measurement is given by HEATH (1970). PARKER (1978) covers many of the necessary statistical techniques, including tests of significance, analysis of variance, correlation and regression.

2.1.2 With a view to computation

It goes without saying that the analytical and statistical techniques to be applied should be firmly in the experimenter's mind before practical work is begun. Looking round for some way of 'doing the stats.' after data capture is complete is a risky process since the value of a lengthy experiment can dissolve in seconds with the realization that the hard-won data are not in a form amenable to analysis. Plan the *whole* experiment first. If this is done not only is the most efficient use made of time and manpower but the job is actually easier since a smooth flow of data can be arranged all the way from the bench to the write-up.

Over and above these considerations, in experiments involving most forms of plant growth analysis the experimenter has to decide which of two approaches to take:

(i) the 'classical' approach, in which the course of events is followed through a series of relatively infrequent, large harvests (with much replication of measurements);

(ii) the 'functional' approach in which the harvests are smaller (less replication of measurements) but more frequent.

The two approaches are not mutually exclusive if time and space are no object (harvests may be large *and* frequent) but it is not often that such a scheme makes the most efficient use of the material available. The classical approach is covered in Chapters 3 and 4 and the functional approach, which depends more or less on some form of computing support, is dealt with in Chapter 5 and, to a large extent, in Chapter 6.

2.2 Harvesting techniques

The stock-in-trade of plant growth analysis is a collection of simple basic data, the measured quantities upon which subsequent analyses depend. These may be determined either for the whole plant or for different sections such as roots, stems and leaves, as required. For *fresh weight*, it is important to maintain standard moisture conditions before and during measurement. For *dry weight*, under- and over drying must be avoided. A reliable system for measuring *length* is not difficult to devise but *volume* must be determined either by calculation, if the geometry of the plant is simple, or, if not, by the displacement of water (employing the Archimedes' principle). *Area*, either of leaves alone or of leaves plus other green parts, is useful and may be determined by several methods: by tracing on to graph paper, then counting squares; by tracing on to ordinary paper or contact-printing on to photographic, blueprint or dyeline paper, then cutting out and weighing the impressions; by planimetry of prints or tracings; or by using automatic machines in which photocells receive light from a standard source in inverse proportion to the area of leaf impressions or other material interposed between the photocells and the source. Readers who are not familiar with reliable methods of determining the quantities they require are referred to MILNER and HUGHES (1968) where relevant details and many useful references are to be found.

A comprehensive review of laboratory techniques for determining *mineral nutrient* contents within plant material, and *organic constituents* such as carbohydrates and proteins, has been provided by ALLEN (1974).

2.3 Units

By international agreement all scientific work is now conducted using the SI system of units (*Système International d'Unités*). The standard unit of length is the metre and of mass the kilogramme. Multiples or fractions of units are restricted to steps of one thousand. For plant growth analysis,

this means that we have available millimetres, metres and kilometres for
length (and hence area) and microgrammes, milligrammes, grammes and
kilogrammes for mass. The very useful centimetre, square centimetre and
square decimetre have been sacrificed to a good cause. The SI units of
energy (joule), power (watt) and customary temperature (degree Celsius)
can be adopted with few, if any, problems. The SI unit of time is the
second. In plant growth analysis, as in some other fields, the processes
studied operate on a longer time scale than this and the everyday units, of
which days and weeks are by far the most useful to us, have been retained.

2.4   Notation

In any field of scientific study a consistent notation is a great advantage,
especially where equations and mathematical expressions abound. Plant
growth analysis has had such a fragmentary evolution that it was not until
EVANS (1972) that a comprehensive and coherent system of notation was
attempted. With unimportant variations this is the system that will be
followed in this booklet. Its conventions are

Contractions of names   :   small capitals; e.g. RGR, relative growth rate
Measured quantities   :   italic capitals; e.g. W, dry weight
Derived quantities   :   bold capitals; e.g. $\mathbf{R}$, relative growth rate
Parameters of equations :   italic lower case; e.g. constants a, b
Distinguishing subscripts: suffix position; e.g. R_W, root dry weight
Subscripts defining time :   prefix position; e.g. $_1T, _1W$, initial time and
                                              dry weight.

By way of further example, all of these conventions will have been
introduced and used by the end of the next chapter. The appendix
contains a synopsis of contractions, symbols, expressions, formulae and
units. An important point to note is that the measured quantity, time,
receives the symbol T and not the more usual t.

3 Growth Analysis of Individual Plants

3.1 Relative and absolute growth rates

Two plants have been grown for a week in an experiment. One weighed 1 g at the beginning and one weighed 10 g. At the end of the week it was found that each had increased its weight by 1 g. Which had grown faster?

From one point of view the performances of the two plants are identical since equal amounts of weight have been gained over equal periods of time; in fact, both plants show the same absolute growth rate, 1 g week^{-1}. But, armed with the knowledge that their initial weights were so dissimilar it is easy to see that the performance of the lighter plant, which doubled its weight, is in an important sense superior to that of the heavier which increased its weight by only a tenth. Given similar performances during the succeeding weeks, the weight of the heavier plant would soon be equalled by that of the other, which initially was ten times the lighter. Clearly, some measure of growth is needed which takes account of this original difference in size.

In the financial world, investors, at least in the short term, would inspect the rates of interest earned, not the amounts of capital held, in order to compare financial skills. The measure of plant growth which is analogous to this rate of interest earned is the relative growth rate, RGR. This is the increase in plant material per unit of material per unit of time (see equation 3/1). In the case of the imaginary experiment described above, the mean relative growth rate of the 1 g plant is 0.69 g g^{-1} week^{-1} and that of the 10 g plant is 0.10 g g^{-1} week^{-1}. At these rates of growth both plants would achieve weights of around 14 g after rather less than four weeks' growth. Obviously, these figures provide a more informative comparison of the plants' relative performances. Where did these methods of comparison originate and how are they executed?

3.2 Relative growth rate

3.2.1 Historical background

The direct analogy with financial investment was developed, in the main, by BLACKMAN (1919). In dealing with plant growth he proposed that the rate of interest be termed the 'efficiency index of dry weight production'. This, Blackman held, is 'clearly a very important physiological constant. *It represents the efficiency of the plant as a producer of new material*, and gives a measure of the plant's economy in working . . . since not only does it indicate the plant's growth efficiency as measured by increase of dry material but it also appears as an exponential term in the equation which expresses the relation between initial dry weight, the final dry weight, and the period of growth' (Blackman's italics). This equation of Blackman's may be written

$$_2W = {_1}W e^{\mathbf{R}(_2T - {_1}T)} \tag{3/1}$$

where $_1W$ is the dry weight at time $_1T$, $_2W$ the dry weight at time $_2T$ and e the base of natural logarithms. The two exponents which appear on the right-hand side of the equation are $\mathbf{R}$ (the 'efficiency index') and the time interval itself, $_2T - {_1}T$. The expression, in effect, involves a mean value of $\mathbf{R}$ for this time interval.

WEST, BRIGGS and KIDD (1920) suggested the name 'relative growth rate' for $\mathbf{R}$ and BRIGGS, KIDD and WEST (1920a) warned that it was dangerous to regard it as a 'constant' for the plant since it showed marked variations in value at different stages of growth. Contributing to this argument from a mathematical point of view, FISHER (1921) pointed out that $\mathbf{R}$ is most simply expressed as an instantaneous value. In calculus notation this reads

$$\mathbf{R} = \frac{1}{W} \cdot \frac{dW}{dT}. \tag{3/2}$$

Another name for $\mathbf{R}$ is 'specific growth rate'. This term is better, in the sense that it is more in line with modern nomenclature, but it is more recent and much less widely used.

3.2.2   Theoretical aspects

The two concepts of instantaneous relative growth rate in weight in plants and its mean over a stated period are analogous to a number of more familiar cases. For example, two ways of estimating the speed of a motor car during a journey would be, firstly, to take the speedometer reading (giving an instantaneous value which would quite possibly be changing continuously) and, secondly, to calculate the mean speed over the whole journey from a knowledge of the total distance travelled and the time taken. As in plant growth analysis, both concepts are valid, each useful for its different purposes.

Firstly, Fisher's expression for instantaneous relative growth rate (equation 3/2) is an exact notation of the definition given in section 3.1 namely, the increase in plant weight per unit of plant weight per unit of time. But, although simple in concept, it is not accessible to direct evaluation in a single plant since, according to the rules of calculus (see MACHIN, 1976, p. 73; CAUSTON, 1977, p. 190) it is equivalent to

$$\mathbf{R} = \frac{d(\log_e W)}{dT}. \tag{3/3}$$

This expression tells us that instantaneous relative growth rate, **R**, is the slope of the plot of the natural logarithms of W against T. Most importantly, its value is free to change with different values of T.

In contrast, mean relative growth rate, $\bar{\mathbf{R}}$, is a more involved concept in theoretical terms but a simpler proposition in practice. Mathematically-minded readers will see that it is derived from the integration of equation 3/2 between the limits $_1T$ and $_2T$, which leads to:

$$\log_e \frac{_2W}{_1W} = \bar{\mathbf{R}}(_2T - {_1T}). \qquad (3/4)$$

This may be re-arranged as

$$_{1-2}\bar{\mathbf{R}} = \frac{\log_e {_2W} - \log_e {_1W}}{_2T - {_1T}}, \qquad (3/5)$$

and from here the mean value, $_{1-2}\bar{\mathbf{R}}$, may be derived simply by substituting-in experimental values of W. As in the case of the car journey, if the amount of change during a given period of time is known then the mean rate of change during this period may be derived.

In practice, the instantaneous values of **R** (equation 3/3) often change smoothly with time and their drift may be followed approximately by deriving values of $\bar{\mathbf{R}}$ for successive harvest intervals via equation 3/5. If the harvest intervals, like $_2T - {_1T}$, are long, $\bar{\mathbf{R}}$ follows **R** only crudely but as the intervals become shorter so the correspondence between these two estimates becomes progressively closer.

3.2.3 Practical approaches

It is clear from equations 3/3 and 3/5 that a plot of the progression of the natural logarithms of weight on time can provide the basis for the calculation of both **R** and $\bar{\mathbf{R}}$. These relative growth rates are both slopes of such a plot, the one instantaneous, the other a mean over a stated period. Figure 3–1 shows this relationship diagrammatically for a case where $\log_e W$ increases non-linearly with T.

The mean RGR over the period $_1T$ to $_2T$, $_{1-2}\bar{\mathbf{R}}$, is the tangent of the mean angle of slope, A. That is

$$_{1-2}\bar{\mathbf{R}} = \frac{\log_e {_2W} - \log_e {_1W}}{_2T - {_1T}} = \frac{y}{x}. \qquad (3/5)$$

In the centre of this time interval the instantaneous RGR, which, to be consistent in notation, should be called $_{1.5}\mathbf{R}$, is similar in value to this $\bar{\mathbf{R}}$ since the tangent to the curve at this point has virtually the same slope in this case as the straight line joining points $_1T$ and $_2T$. However, **R** is an instantaneous slope (the tangent to the curve at each point in question) and at any time its value is

$$\mathbf{R} = \frac{d(\log_e W)}{dT} \qquad (3/3)$$

So, at $_1T$ the instantaneous relative growth rate, $_1\mathbf{R}$, will clearly be greater than $\bar{\mathbf{R}}$. Similarly, at $_2T$ the corresponding instantaneous relative growth rate, $_2\mathbf{R}$, will be less than $\bar{\mathbf{R}}$.

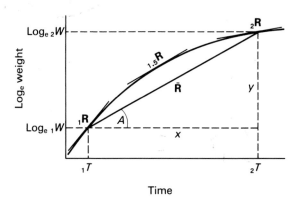

Fig. 3–1   The derivation of instantaneous and mean relative growth rates from a plot of $\log_e W$ against T.

In the special case of exponential growth the progression of $\log_e W$ on T is linear throughout. All values of $\mathbf{R}$ are the same and are equal to $\bar{\mathbf{R}}$ wherever they may occur in time.

The derivation of $\mathbf{R}$ will be discussed further in Chapter 5. For the moment, we can use equation 3/5 to calculate values of $\bar{\mathbf{R}}$ for the imaginary example given in section 3.1. The different stages involved in these calculations are laid out in Table 2.

Table 2   Calculation of mean relative growth rate, $_{1-2}\bar{\mathbf{R}}$, using equation 3/5 in the case of the imaginary example given in section 3.1.

Quantity	Smaller plant	Larger plant
$_1W$	1 g	10 g
$_2W$	2 g	11 g
$\log_e {_2}W - \log_e {_1}W$	0.693 − 0	2.40 − 2.30
$_2T - {_1}T$	1 week	1 week
$_{1-2}\bar{\mathbf{R}}\ (\mathrm{g\,g^{-1}\,week^{-1}})$	0.693	0.10

The units of $\bar{\mathbf{R}}$ in this case could more economically be expressed as 'week^{-1}', since the two weight components of the unit cancel into a dimensionless fraction. Of course, any units of the form size size^{-1} time^{-1}

may be used but there are certain advantages attached to each of the more commonly-used versions:

$[g\,g^{-1}]\,week^{-1}$ – the time unit is often the same as the harvest interval and values tend to be of a convenient size;

$[g\,g^{-1}]\,day^{-1}$ – convenient when T is measured in days, but values tend to be rather low;

$mg\,g^{-1}\,week^{-1}$ – considerable accuracy is available without the need for decimal places.

In addition, it may be mentioned that per cent per week, or per day have also been used where easy interpretation is required across a large range of values (as in Table 1).

In the imaginary example given it is not specified whether 'weight' is fresh weight or dry weight. If the latter, then destructive harvesting would have been needed to estimate both $_1W$ and $_2W$. This subsequently raises problems on the extent to which estimates of $\bar{R}$ truthfully represent the population means for these quantities. Repeated sampling for W is necessary among sub-samples of each population and some estimate of variability must be derived for $\bar{R}$. In the example given, the values of $\bar{R}$ in the two populations are probably distinct enough not to need statistical analysis unless the replicated values of W were highly variable. But, in the more general case, an exact test is required. This is done by calculating a series of $\bar{R}$s for paired plants: the largest at time 1 paired with the largest at time 2; and second largest at time 1 and the second largest at time 2; and so on. This process requires equal replication at both harvests. Alternatively, plants at each harvest may be grouped into large, medium and small, leading to just three estimates of $\bar{R}$ using mean values, $\bar{W}$, for each group. Obviously, both the number and variability of these estimates critically affects the significance of the overall $\bar{R}$ which may then be derived from these intermediate data. Replicated measurements like these make it possible to fix a standard error (PARKER, 1978, § 2.4) to the overall value of $\bar{R}$ if the calculations laid out in Table 3 are repeated for each plant, or group of plants, paired across the harvest interval.

Finally, the absolute growth rate, G, of the plants in the example given in section 3.1 is

$$G = \frac{dW}{dT}.$$ (3/6)

This G is the instantaneous slope of the plot of W against T, a plain and simple measure of the rate of dry weight increase. Its mean value, $\bar{G}$, over the time interval in question is given by

$$_{1-2}\bar{G} = \frac{_2W - _1W}{_2T - _1T}.$$ (3/7)

As we have seen, $\bar{G}$ in both plants had a value of 1 g week^{-1}.

3.2.4   Effects of ontogeny and environment

Since RGR is equivalent to the slope of the plot of the *logarithms* of W against T (Fig. 3–1) an inspection of such plots will always give a reliable first impression of the behaviour of RGR without the need for calculation at this stage. For example, in the case of *Zea mays* (maize) grown at Poppelsdorf in 1878 it seems from Fig. 1–1b that this slope is initially zero, or even negative, it increases rapidly to a maximum near the middle of the growing period, and then gradually declines (perhaps to zero) at the end. When values of mean RGR are calculated on a weekly basis, using equation 3/5, these trends are confirmed (Fig. 3–2). An initial fall to below zero soon recovers; RGR reaches a maximum in the sixth week and then drops to zero (with some large fluctuations) in the sixteenth week.

Most of this change is an expression of ontogenetic drift. This is due to developments which occur within the plant with the passing of time. But these developments occur against the background of a changing environment (the plants were grown outdoors over the period May to September) and in this case it is impossible to disentangle these internal and external influences on RGR. When plants are grown in a constant, controlled environment for a lengthy period a truer picture of ontogenetic drift in RGR emerges (Fig. 3–3). In this example the earliest phases of growth are not represented since measurements began only at the time of maxiumum RGR. Also the curve can be made smoother than that of Fig. 3–2 because in this experiment random fluctuations in environmental conditions were absent.

It is impossible in a booklet of this length to give a complete list of the effects on RGR of all the environmental variables that have at one time or another been examined. In general, any departure from an adequate supply of light, mineral nutrients or water, or from a suitable temperature regime, or from freedom from external toxins, produces a clearly adverse effect on RGR. Such is the sensitive linkage of this quantity to the whole environmental relations of the plant. It might be added that these factors also interact strongly. For example, the plant's growth response to 'low' levels of one factor depends very much on the available levels of the other factors (see Table 7 for an example of this). The fullest examination yet made of the effects of the environment on the relative growth rate (and other features) of a single species under comparable conditions is that of G. C. Evans and A. P. Hughes working with *Impatiens parviflora* (small balsam). This is summarized by HUGHES (1965).

3.2.5   Differences between species

The first comparison of inter-specific differences in RGR also came from Blackman who recalculated previously-published data on the growth of young crop plants (Table 3). The subsequent literature contains values of $\mathbf{R}$, or $\bar{\mathbf{R}}$, for several hundred species, but since relatively few of these were

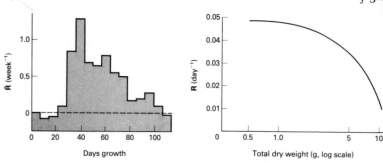

Fig. 3-2 Changes in mean relative growth rate, $\bar{R}$, with time in Badischer Früh maize grown at Poppelsdorf in 1878. Recalculated from data given by KREUSLER, PREHN and HORNBERGER, 1879. See also Fig. 1-1.)

Fig. 3-3 Ontogenetic drift in instantaneous relative growth rate, R, in *Chrysanthemum moriflorium* cv. Bright Golden Anne grown in a productive, controlled environment. (From HUGHES, 1973.)

grown under strictly comparable conditions only a limited number of comparisons of inter-specific differences is possible. So far, the largest body of comparable data available here is that of GRIME and HUNT (1975). One hundred and thirty-two species (mainly native to Britain) were grown under favourable controlled conditions and their growth was

Table 3 Estimates of mean relative growth rate in young crop plants grown under favourable conditions. (From BLACKMAN, 1919.)

Species	$\bar{R}$ (day^{-1})
Cannabis sativa cv. *gigantea* (hemp)	0.13
Helianthus macrophyllus giganteus (sunflower)	0.18
Nicotiana tabacum (tobacco)	0.21
Zea mays (maize)	0.07

analysed over the period two to five weeks after germination. The maximum instantaneous RGR, R_{max}, observed for each species during this period ranged from 0.22 week^{-1} (in *Picea sitchensis*, Sitka spruce seedlings) to 2.70 week^{-1} (in *Poa annua*, annual poa). The distributions of Grime and Hunt's small samples of woody and annual species were respectively low- and high-biased in R_{max}. Grasses and forbs (herbs that are not grasses) both included a wide range of growth rates.

Using local field survey data, Grime and Hunt were able to show that in several disturbed and/or productive habitats around Sheffield, fast-growing species were predominant and species of low growth potential were virtually or completely absent. The reverse was true of several of the more stable, unproductive habitats and species of moderate growth potential were ubiquitous.

3.2.6   Differences within species

From the amount of information available (again limited) it seems possible that intra-specific differences in RGR may occasionally rival the magnitude of inter-specific differences, at least within ecologically-similar groups of plants. Within crop species, of course, much comparative information is available from inter-varietal trials, e.g. DUNCAN and HESKETH (1968), but in British species one of the largest intra-specific comparisons of RGR has been that of A. J. M. Baker (unpublished) working at Sheffield. Seven distinct local collections of *Holcus lanatus* (Yorkshire fog) and five of *Festuca ovina* (sheep's fescue) were grown from seed under productive, solution-culture conditions. Differences in $\bar{R}$ between collections of the same species were slight, but occasionally significant at $P < 0.05$ (Table 4).

Table 4   Intra-specific comparisons of mean relative growth rate in *Holcus lanatus* and *Festuca ovina*. Values given cover the period 2–7 weeks after germination. (Unpublished data of A. J. M. Baker.)

Species and source	Soil pH	$\bar{R}$ (day^{-1})	±95% limit
Holcus lanatus			
Bingham Park	4.5	0.191	0.008
Lathkilldale (plateau)	4.5	0.196	0.007
Lathkilldale (south-slope)	7.2	0.200	0.008
Lodge Moor	7.3	0.188	0.006
Loxley	4.1	0.191	0.007
Warsop Vale	5.0	0.195	0.006
Whiteley Woods	4.2	0.183	0.008
Festuca ovina			
Burbage	3.6	0.154	0.012
Lathkilldale (south-slope)	7.2	0.130	0.009
Loxley	3.7	0.149	0.012
Winnats Pass (north-slope)	7.1	0.137	0.010
Winnats Pass (south-slope)	7.4	0.152	0.013

On a broader geographical scale, EAGLES (1969) has shown that in a productive, controlled environment there can often be a two-fold difference in **R** between Norwegian and Portuguese populations of *Dactylis glomerata* (cocksfoot) although both the direction and magnitude of this differential are subject to marked interactions of ontogeny and environment.

3.2.7   Usefulness

In conclusion, it may be said that relative growth rate provides a convenient integration of the combined performances of the various parts of the plant. It is especially useful when the need arises to compare species and treatment differences on a uniform basis. But when calculated at the whole plant level it tells us nothing of the causal processes which contribute to the plant's gross performance. One further step in this direction is possible before going on to examine other ways of assessing the growth of the plant. This is to calculate RGR for each sub-component of the plant, say root and shoot; or leaves, stem, and root; or leaves of varying ages, petioles, stem, main and lateral roots. The subdivisions are dictated only by convenience and the computations are directly analogous to those for whole plant RGR. Similarly, important new information may be gained from RGR calculations made on the basis of fresh weight, volume, area, or length; or on the basis of the content of various metabolic compounds such as selected carbohydrates or proteins, or even the calorific content of the plant material. In this wider context the term 'relative growth rate' is to be preferred to 'specific growth rate' (section 3.2.1) since the latter refers exclusively to growth in relation to mass.

3.3   Unit leaf rate and leaf area ratio

3.3.1   Historical background

The expression for instantaneous RGR,

$$\mathbf{R} = \frac{1}{W} \cdot \frac{\mathrm{d}W}{\mathrm{d}T},$$ (3/2)

treats all of the weight of the plant as being equally productive of further weight. We know that as plants grow the proportion of purely structural material that they contain increases, for much the same reasons that larger animals develop proportionately more bulky bones than smaller ones (ALEXANDER, 1971, p. 53). So this implied notion of mere weight being all that is necessary to provide still more weight becomes more and more improbable as growth proceeds. What is needed is an index of the productive efficiency of plants in relation to some clearly-identifiable component of relatively constant performance. GREGORY (1918) suggested that the net gain in weight per unit of leaf area (the 'average rate of assimilation') might be this more meaningful index of growth. BRIGGS, KIDD and WEST (1920b) termed it 'unit leaf rate' (ULR). This quantity is conventionally given the symbol **E** and the expression for its instantaneous value is

$$\mathbf{E} = \frac{1}{L_\mathrm{A}} \cdot \frac{\mathrm{d}W}{\mathrm{d}T}$$ (3/8)

where L_A is the total leaf area present on the plant. WILLIAMS (1946) provided a convenient formula for the estimation of mean ULR, $\bar{E}$, over a period of time:

$$_{1-2}\bar{E} = \frac{_2W - _1W}{_2T - _1T} \cdot \frac{\log_e {_2L_A} - \log_e {_1L_A}}{_2L_A - _1L_A}. \tag{3/9}$$

Unit leaf rate has also widely been called 'net assimilation rate', NAR, (GREGORY, 1926) but the former term is both older and more suitable, as EVANS (1972, p. 205) explains.

Armed with what might, but for the inclusion of the weight of mineral elements in W, be an estimate of the carbon-assimilatory capacity of the leaves, all one now lacks is an estimate of the leafiness of the plant before being in a position to calculate the overall relative growth rate. Alternatively, beginning with RGR, this index of leafiness is the other quantity that can be derived, with ULR, to produce an informatively subdivided summary of the plant's performance. BRIGGS, KIDD and WEST (1920b) called this other quantity the leaf area ratio (LAR) and defined it as the ratio of total leaf area to whole plant dry weight. It can be notated as $\mathbf{F}$:

$$\mathbf{F} = \frac{L_A}{W}. \tag{3/10}$$

In a broad sense, LAR represents the ratio of photosynthesizing to respiring material within the plant.

Over a harvest interval its mean value, $\bar{\mathbf{F}}$, is simply given by

$$_{1-2}\bar{\mathbf{F}} = \frac{(_1L_A/_1W) + (_2L_A/_2W)}{2} = \frac{_1\mathbf{F} + _2\mathbf{F}}{2} \tag{3/11}$$

if one assumes that $\mathbf{F}$ is linearly related to time (ONDOK, 1971, discusses other cases). Equation 3/10 supplies a straightforward estimate of instantaneous value using corresponding values of L_A and W from a single harvest.

Since ULR and LAR evolved simultaneously as subdivisions of RGR it is by definition that

$$RGR = ULR \times LAR \tag{3/12}$$

or

$$\mathbf{R} = \mathbf{E} \times \mathbf{F} \tag{3/13}$$

or

$$\frac{1}{W} \cdot \frac{dW}{dT} = \frac{1}{L_A} \cdot \frac{dW}{dT} \times \frac{L_A}{W}. \tag{3/14}$$

Simply expressed, the growth rate of the plant depends simultaneously upon the efficiency of its leaves as producers of new material and upon the leafiness of the plant itself. But, except in very special circumstances,

$$\bar{\mathbf{R}} \neq \bar{\mathbf{E}} \times \bar{\mathbf{F}} \tag{3/15}$$

because equation 3/13 holds only crudely for mean values of the three quantities. Instantaneous values are needed for this relationship to be precise.

3.3.2   Theory and practice

The expression for mean ULR (equation 3/9) makes the assumption that weight and leaf area are linearly related over the period of observations. That is,

$$W = a + bL_A \qquad (3/16)$$

where a and b are constant for the case in hand. Users should be satisfied that equation 3/16 is an accurate summary of this relationship before employing equation 3/9. The two conditions in which, above all, equation 3/16 is likely to be untrue are (i) when plants are growing quickly, and (ii) when the harvest intervals are long. If the problem cannot be solved by (or if it is too late for) more frequent harvesting, other methods have to be used.

Various formulae are available for these other cases where W varies as some more complex function of L_A. But in the classical approach to plant growth analysis the best method for dealing with non-linearity between W and L_A has been put forward by EVANS and HUGHES (1962) and elaborated by EVANS (1972, p. 268). Here, the more general case of

$$W = a + bL_A{}^n \qquad (3/17)$$

is considered. Knowing both n (power) and the ratio ${}_2L_A/{}_1L_A$ the user may refer to tables given by EVANS (1972, Appendix 2) to find the percentage difference between $\bar{E}$ values calculated on the assumption of 3/17 and the corresponding value for the assumption that

$$W = a + bL_A{}^2 \qquad (3/18)$$

for which the expression for $\bar{E}$ is simply

$$_{1-2}\bar{E} = \frac{{}_2W - {}_1W}{{}_2T - {}_1T} \cdot \frac{2}{{}_2L_A + {}_1L_A}. \qquad (3/19)$$

After the best approach for computation has been chosen the errors of $\bar{E}$ may be derived in the same way as those of $\bar{R}$ (section 3.2.3). The same is true for those of $\bar{F}$. At this point it is worth mentioning that if one has the facilities to compute instantaneous values of E (Chapter 5) many of these problems concerning the correctness or not of assumptions and approximations are evaded.

3.3.3   Effects of ontogeny and environment

One of the factors that led to the introduction of unit leaf rate was a search for a relatively constant index of growth that was independent of plant size. Although, in general, ULR proves to be stable for longer periods than RGR its ontogenetic drift is still marked. WILLIAMS (1946) showed that for annual plants grown in a constant environment, the closer the approach to an effective measure of assimilating capacity, the more reliable and characteristic of the species become the estimates of $\bar{E}$ derived from this measure. When calculated on the bases of leaf weight, leaf area and leaf protein a clear series of increasing stability of $\bar{E}$ emerged, particularly when nitrogen was in relatively short supply. These three versions of $\bar{E}$ may be notated $\bar{E}_W$, $\bar{E}_A$ and $\bar{E}_P$ respectively.

In cases where it is not feasible to estimate $\bar{E}_P$, $\bar{E}_A$ can usually provide an acceptable substitute.

In a simultaneous comparison of the growth of *Hordeum vulgare* (barley) outdoors at Ottawa and in a controlled environment, THORNE (1961) showed that a downward ontogenetic drift in $\bar{E}_A$ was accentuated by the declining favourability of outdoor conditions in the late summer (Fig. 3–4).

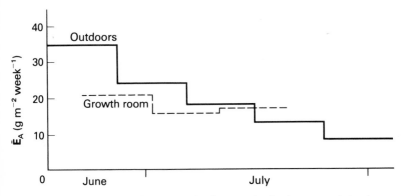

Fig. 3–4 A comparison of mean unit leaf rates in barley (var. Brant). (Redrawn from THORNE, 1961.)

A striking example of this dependence of $\bar{E}_A$ on the external environment was provided by the work of WATSON (1947). Each of four species, *Beta vulgaris* (sugar beet), *Hordeum vulgare* (barley), *Solanum tuberosum* (potato) and *Triticum* sp. (wheat) grown in the field at Rothamsted showed a peak in $\bar{E}_A$ near the Summer Solstice (late June) irrespective of the nature of the crop or of the time of planting (Fig. 3–5).

These peaks are principally the result of a combination of high solar insolation, high temperature and long day-length. Each of these factors may be investigated singly under controlled or semi-controlled conditions. For example, BLACKMAN and WILSON (1951) used a series of shade screens to demonstrate the dependence of $\bar{R}$, $\bar{E}_A$ and $\bar{F}$ upon the level of illumination received. In general, $\bar{E}_A$ and $\bar{F}$ were found to be

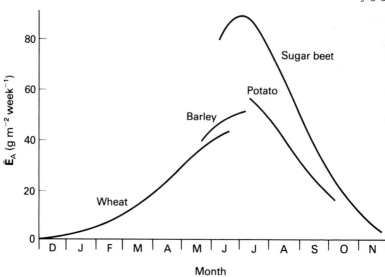

Fig. 3–5   Smoothed seasonal changes in mean unit leaf rate in outdoor crops. (From WATSON, 1947.)

related linearly to the logarithm of percentage light intensity; in $\bar{F}$ this relationship was an inverse one. The slopes of these two relationships determined the trend obtained for $\bar{R}$. In one extreme case, *Geum urbanum* (wood avens), maximum $\bar{R}$ was predicted to occur at only 54% of full daylight because of the steep plunge in $\bar{F}$ that occurred with increasing light intensity (Fig. 3–6).

3.3.4   *Differences between species*

In an early examination of their own and other workers' data on inter-specific differences in mean unit leaf rate, HEATH and GREGORY (1938) found almost six-fold variation in $\bar{E}_A$ among British crops grown in summer. Values ranged from 12.5 g m^{-2} week^{-1} in *Cucurbita pepo* (cucumber) to 72.0 g m^{-2} week^{-1} in *Beta vulgaris* (sugar beet).

It is now confirmed that wide variation in $\bar{E}_A$ may occur between species. For example, unpublished data from the experiments of GRIME and HUNT (1975) show that values of E_A (instantaneous maxima, E_{max}) vary from 19.7 g m^{-2} week^{-1} (*Vaccinium vitis-idaea*, cowberry) to 192 g m^{-2} week^{-1} (*Cerastium holosteoides*, mouse-ear chickweed) in their particular sample of species. Dicotyledons, monocotyledons and annual species show no distinct bias towards high or low E_{max} but woody species are clearly low biased.

COOMBE (1960) concluded that woody plants exhibited inherently lower E_A than the majority of herbaceous species, a conclusion supported by the work of JARVIS and JARVIS (1964). These workers grew a variety of coniferous species and sunflower under productive, controlled conditions. They showed that in comparison with sunflower, the lower $\bar{R}$ of the conifer seedlings was due more to a low $\bar{F}$ than to a low $\bar{E}_A$ (although there were important differences in $\bar{E}_A$ also). Table 5 gives a typical comparison (equation 3/15 is in operation here).

Table 5 Contribution of $\bar{E}_A$ and $\bar{F}$ to seedling $\bar{R}$. (Recalculated from data given by JARVIS and JARVIS, 1964.)

Species and harvest interval	$\bar{R}$ (week^{-1})	$\bar{E}_A$ (g m^{-2} week^{-1})	$\bar{F}$ (m^2 g^{-1})
Pinus sylvestris (Scots pine) 42 days	0.135	32.3	0.0044
Helianthus annuus (sunflower) 8 days	0.955	59.3	0.0177

3.4 Specific leaf area and leaf weight ratio

3.4.1 Leaf area ratio subdivided

If leaf dry weight, L_W, is known a useful subdivision of LAR may be made:

$$\frac{L_A}{W} = \frac{L_A}{L_W} \times \frac{L_W}{W} \qquad (3/20)$$

or

$$\text{Leaf area ratio} = \text{Specific leaf area} \times \text{Leaf weight ratio} \qquad (3/21)$$
$$\text{(LAR)} \qquad\qquad \text{(SLA)} \qquad\qquad \text{(LWR)}$$

SLA is the mean area of leaf displayed per unit of leaf weight (in a sense a measure of leaf density or relative thickness) and LWR is an index of the leafiness of the plant on a weight basis (cf. SLA, the leafiness of the plant on an area/weight basis). These subdivisions of LAR may be inserted into equation 3/14 to give:

$$\frac{1}{W} \cdot \frac{dW}{dT} = \frac{1}{L_A} \cdot \frac{dW}{dT} \times \frac{L_A}{L_W} \times \frac{L_W}{W} \qquad (3/22)$$

or

$$\text{RGR} = \text{ULR} \times \text{SLA} \times \text{LWR} \qquad (3/23)$$

As with LAR, SLA and LWR are both amenable to calculation of instantaneous values at the time of harvest. Mean values between harvest intervals may also be estimated in the same manner as those of LAR (equation 3/11) and statistical analyses may be performed as before (section 3.2.3).

3.4.2   *Effects of ontogeny and environment*

Of the two, SLA and LWR, the former is in general both the more sensitive to environmental change and the more prone to ontogenetic drift. Referring to the growth of *Impatiens parviflora* (small balsam) EVANS (1972, p. 331) wrote: 'For leaf weight ratio we could draw up a list of environmental factors, changes in which hardly affected the value of LWR. It would be profitless to attempt to do the same for specific leaf area, as there has been at least some influence of the environment in every instance which we have examined.' Foremost in this second list might be variation in light intensity: deep shade causes striking increases in SLA which may partly offset decreases in ULR (HUGHES and EVANS, 1962).

3.4.3   *Differences between species*

Differences in both LWR and SLA occur even between closely-related species. For example, EVANS (1972, p. 442) explains that in two species of sunflower, *Helianthus debilis* 'has a substantially higher proportion of its dry matter in the form of leaves than has *H. annuus*'; values of LWR were commonly greater by c. 20%.

JARVIS and JARVIS (1964) established that the much less leafy nature of Scots pine, in comparison with sunflower (Table 5), was due almost entirely to the relatively greater density of the pine needles and hardly at all to any variation in LWR (the 'productive investment' of the plant) which, in fact, showed a small difference in favour of pine (Table 6).

Table 6   Contributions of SLA and LWR to seedling LAR (instantaneous values). (Recalculated from data given by JARVIS and JARVIS, 1964.)

Species and harvest	LAR $(m^2 g^{-1})$	SLA $(m^2 g^{-1})$	LWR
Pinus sylvestris, Scots pine (at 2 years)	0.0054	0.0084	0.643
Helianthus annuus, sunflower (at c. 10 cm height)	0.0234	0.0432	0.542

3.5   Other ratios

Many other ratios between two simple measurements have been constructed. A selection of the more frequently used is

$$\text{Root/shoot ratio,} \frac{R_W}{S_W};$$

Root (or shoot) weight fraction (or ratio), $\frac{R_W}{W}$, $\frac{S_W}{W}$;

Fresh weight/dry weight ratio, $\frac{FW}{W}$;

where R_W and S_W are root and shoot dry weights respectively and FW is whole plant fresh (wet) weight. Direct calculations from harvest data are feasible in each case and mean values over a harvest interval may be estimated using methods analogous to those for $\bar{F}$ (equation 3/11). All ratios are subject to genetic, ontogenetic and environmental control.

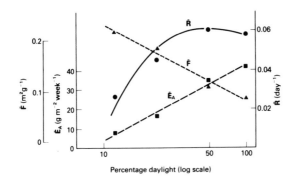

Fig. 3–6   The dependence of mean relative growth rate, $\bar{R}$, mean unit leaf rate $\bar{E}_A$, and mean leaf area ratio, $\bar{F}$, in *Geum urbanum* (wood avens) on the logarithm of percentage daylight. (From BLACKMAN and WILSON, 1951.)

3.6   Allometry in plant growth

3.6.1   Historical background

Allometry in biology involves the study of the growth and development of one part of an organism in relation to another. The main developments of this topic in plant science have been due to PEARSALL (1927) and TROUGHTON (1955). An excellent introduction to allometry in animal science is given by ALEXANDER (1971, p. 1).

3.6.2   Theory and practice

In general terms, the allometric relationship between two plant variates X and Y may be expressed thus:

$$Y = bX^{K} \tag{3/24}$$

where X represents some parameter of the size of the whole organism (or some distinct part of it) and Y represents the same for some differentially-growing component of it; b and K are both constant for the case in hand. Normally K is called the 'allometric constant'.

Taking logarithms of equation (3/24);

$$\log Y = \log b + \mathbf{K} \log X. \qquad (3/25)$$

This provides a convenient practical method of evaluating equation 3/24 since a plot of Y against X on a double logarithmic scale will feature $\mathbf{K}$ as its slope. Alternatively, $\mathbf{K}$ may be determined with more precision from a linear regression of the logarithms of Y on those of X (see BAILEY (1964, p. 91) or PARKER (1978, § 8.6) for the correct method of doing this).

Because $\mathbf{K}$ can summarize the whole behaviour of X and Y across any number of harvests it is of considerable practical value. Much may be implied with great economy of expression, since the value of $\mathbf{K}$ is, in effect, the ratio of the logarithmic growth rates of the two components under investigation. Any value other than unity implies a discrepancy between these two rates which will lead to a change in the relative proportions of the components as time goes by. Thus ontogentic drift may be described by an index of growth which is itself constant for substantial periods – an unusually advantageous situation.

3.6.3   Allometry in operation

So far, the greatest use made of allometry in plant science has been in the examination of root/shoot relationships. For this purpose equation 3/25 may be re-written

$$\log R_{\mathrm{W}} = \log b + \mathbf{K} \log S_{\mathrm{W}}. \qquad (3/26)$$

When values of $\mathbf{K}$ are greater than unity the plant progresses more and more towards 'rootiness'. When less than unity this progression is towards 'shootiness'. The value of $\mathbf{K}$ is normally constant for the species and environment during the vegetative phase of growth but TROUGHTON (1956) has shown that in *Lolium multiflorum* (Italian ryegrass) it drops abruptly at the time of flowering (Fig. 3–7). Marked genotypic variation in $\mathbf{K}$ also occurs.

The environment has a strong influence on $\mathbf{K}$. For example, when BROUWER, JENNESKENS and BORGGREVE (1961) transferred *Zea mays* (maize) plants that had previously been supplied with nitrogen to a solution without nitrogen, a clear increase in the value of $\mathbf{K}$ occurred.

In general, moisture- or nutrient-stress tends to increase $\mathbf{K}$ and shading tends to decrease it. These effects may interact: HUNT (1975) showed that in the vegetative phase shaded plants of *Lolium perenne* cv. S.23 (perennial ryegrass) had a lower $\mathbf{K}$ value than controls. Plants grown both on the low nitrogen medium and under shade had a $\mathbf{K}$ value similar to that of controls. Although dry weight was much reduced, these plants were almost isometric with the controls (Table 7).

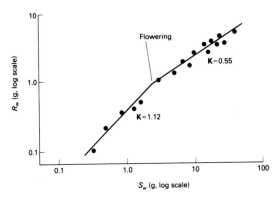

Fig. 3–7   The effect of flowering on the allometric relationship between root and shoot dry weight in *Lolium multiflorum* (Italian ryegrass). (From TROUGHTON, 1956.)

Table 7 Values of K, the allometric constant, for the vegative growth of *Lolium perenne* cv. S.23 (perennial ryegrass) under different conditions. 95% limits are attached. (From HUNT, 1975.)

	Full light	*Shade*
Full nutrients	0.84 ($\pm$0.04)	0.69 ($\pm$0.10)
Low nitrogen	0.95 ($\pm$0.04)	0.73 ($\pm$0.10)

3.6.4   Further allometric relationships

Allometric computations need not be confined either to these two components (root and shoot) or to the use of weight as the index of size. Other measurements such as length, area or volume are quite suitable, the more so if they are non-destructive. Nor need each of the pair of units used be of the same type. For example, WHITEHEAD and MYERSCOUGH (1962) computed what is effectively the allometric relationship between whole plant dry weight and total leaf area:

$$W = a + bL_A^K \qquad (3/27)$$

where **K** is again the allometric constant (called α by Whitehead and Myerscough) and *a*, *b* are also constant for the case in hand. Whitehead and Myerscough described **K** in this instance as indicative of 'the proportion of dry weight increment surplus to that required to maintain the morphogenetic proportions of the plant as an efficient photosynthetic form alone'.

4 Growth Analysis of Populations and Communities

4.1 Links with the growth analysis of individual plants

All of the topics dealt with in the previous chapter concern the growth of plants as spaced individuals. This is not to say that the various concepts were applied only to plants grown singly; far from it, since large populations of similar individuals normally need to be raised to meet the demands of sequential destructive harvesting. Rather that the results, once obtained, were expressed on a 'per plant' basis. The size of the population, and the sum performance of its constituents, were not matters of concern in themselves.

However, in agriculture particularly, and in some studies of production in natural vegetation, it has often been informative to treat both populations (such as monospecific crop stands) and communities (such as mixed-species grassland) as single functional units, expressing their overall performances in terms parallel to those used in the growth analysis of individual plants. It should be emphasized that there is no theoretical reason why the concepts of relative growth rate, unit leaf rate, leaf area ratio and so on, should not themselves be applied on a 'per crop' basis instead of on a 'per individual' basis. But, in practice, this parallel series of analytical procedures has borrowed only one of these, unit leaf rate, for use in an unaltered form. Other concepts, designed exclusively for the study of population and community growth, have been developed to supplement it. After recognizing unit leaf rate in this other role, the present chapter will be concerned mainly with the newer developments.

4.2 Unit leaf rate again

In the examples of the behaviour of unit leaf rate given in section 3.3 no distinction was made between estimates of $\mathbf{E}$ (or $\bar{\mathbf{E}}$) derived from values of L_A and W obtained either from individual plants, or from individuals drawn as sub-samples out of a larger population. When the rate of dry matter production per unit quantity of leaf material is under examination, these two approaches differ neither in theory nor in practice. Hence, we saw examples of the former (Fig. 3–6) given without distinction alongside examples of the latter (Fig. 3–5). In the study of population and community growth, unit leaf rate provides exactly the same information as it does in the study of the growth of individuals, namely, an index of the functional efficiency of the productive parts of the plant. The same formulae for its calculation hold (equations 3/9, 3/19) and the same prerequisites apply (section 3.3.2). In addition, E will henceforward refer to E_A, unless otherwise specified.

4.3   Leaf area index

4.3.1   Historical background

Given that unit leaf rate provides an adequate estimate of the efficiency of unit quantities of the crop's leaves as producers of dry material, then a knowledge also of the leafiness of the crop is needed before its full performance can be assessed. Leaf area per plant is an inappropriate measure of the leafiness of a whole crop since it takes no account of the spacing of the plants, a factor which must clearly be involved in any estimate of 'crop leafiness'. To overcome this difficulty WATSON (1947) introduced the more crop-orientated concept of leafiness in relation to land area. This he named leaf area index, LAI, and defined it as leaf area per unit area of land. Using P to represent the land area and L_A to represent, not the total leaf area per plant as previously, but the total leaf area above the land area P, leaf area index may be given the single symbol L and notated

$$L = \frac{L_A}{P}. \qquad (4/1)$$

LAI is the functional size of the crop standing on the land area P.

If the dimensions of L_A and P are both the same (e.g. area) then L itself has no units: it is a dimensionless ratio. For example, 3.0 m² of leaf area above a land area of 2.0 m² represent a LAI of 1.5. To determine instantaneous values of LAI, L, requires only that L_A be measured in an adequate and representative number of small samples of the crop, each from a known land area, P. Alternatively, L_A may be estimated on a per plant basis and then multiplied by the measured current plant density in the crop. These estimates of L may be subjected to statistical analysis to determine the significance of any differences between species or treatments and, if the time-course of L is to be followed, these measurements may be repeated at intervals. What follows is a brief résumé of the observed behaviour of LAI, considered in isolation. This leads later to a consideration of the overall performance of the crop or stand of vegetation in relation both to its LAI and to its ULR.

4.3.2   Leaf area index in operation

In effect, leaf area index is the number of complete layers of leaves displayed by the crop, expressed as an average for the whole crop. This concept is inevitably a crude one since leaves never actually form complete unbroken layers arranged one above the other. Leaves are commonly displayed at varying angles to the horizontal and these angles vary with the morphology of the species and with the conditions under which it is being grown. LAI is only an index of mean crop leafiness since, even at relatively high values of LAI, random holes occur in the canopy where there is no leaf cover at all. Moreover, leaves at different layers in the canopy experience different environmental conditions and function in different ways. All of these variations are smothered in the simple expression of crop leafiness as L_A/P.

Nevertheless, such variation also lies behind other growth analysis quantities and as a broad index of the productive capacity of a stand of vegetation LAI has been of considerable value. In a newly-germinated crop L remains below 1.0 for some time, since the total leaf area of the young seedlings is negligible in relation to the land area on which they stand. As the crop develops L increases, until it reaches its maximum value (often in the range 2 to 10 for temperate crops). WATSON (1947) presented values of L which, although low by present standards, conveniently illustrate the nature of the seasonal trends in LAI. In *Beta vulgaris* (sugar beet), *Hordeum vulgare* (barley), *Solanum tuberosum* (potato) and *Triticum* sp. (wheat) grown at Rothamsted (Fig. 4–1), L was closely related to the time of planting of the crop, and to its subsequent phenology. It was largely independent of seasonal changes in the environment (cf. $\bar{E}_A$ in the same four species, Fig. 3–5, which is dominated by external conditions). WATSON (1971) has reviewed more recent work in this field.

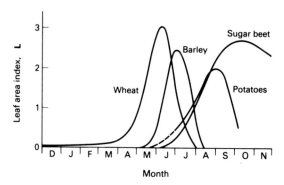

Fig. 4–1   Smoothed seasonal changes in leaf area index in outdoor crops. (From WATSON, 1947.)

In theory, an optimum LAI (defined as that which supports the maximum rate of dry matter increase) is reached when the lowest layers of leaves are only just able, on average, to maintain a positive carbon balance. That is, when the crop as a whole intercepts virtually all of the available photosynthetically-active radiation. Beyond this point such leaves either become an unproductive burden to the crop or plant community, the overall performance of which will begin to fall, or they die. In some crops, however, where total dry weight is not the economic objective, it can be profitable to surpass the optimum LAI if grain, tuber or fruit production benefits at the expense of vegetative growth.

Unlike the other growth analysis quantities so far considered, LAI is to some extent under the direct control of the grower or experimenter. BLACKMAN (1968), quoting unpublished results of J. Newton, showed that LAI in *Gladiolus* (cv. Scheherazade) could be increased to around 27 by planting at a density of 688 plants m^{-2}. Productivity in the whole *Gladiolus* stand increased greatly. However, this increase in LAI was less than proportional to the increase in plant numbers. In an economic sense a balance would have to be struck between increased production and increased outlay and this balance would be over and above the foregoing considerations of 'pay-load' production where this is different from total dry weight. In *Gladiolus* itself, the leaves are held almost vertically. In plants with leaves held more horizontally such high values of LAI are unusual.

Since different environmental conditions have a marked effect on the growth and development of stands of vegetation, these conditions are naturally reflected in the values of LAI observed. For example, BLACK (1963) demonstrated that the optimum LAI for dry matter production in *Trifolium subterraneum* (subterranean clover) was strongly influenced by the level of solar radiation received (Fig. 4–2). The higher this level, the higher the LAI at which maximum production could be sustained. In common with other growth analysis quantities, LAI is also strongly affected by temperature and by the water and mineral nutrient regimes of the crop. WATSON (1952, 1971) provides convenient introductions to these topics.

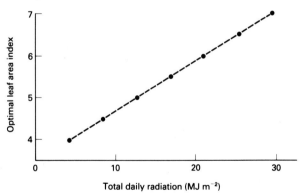

Fig. 4–2   The dependence of optimal leaf area index on total daily radiation in *Trifolium subterraneum*. (Plotted from data given by BLACK, 1963.)

4.3.3   *Leaf area index in relation to unit leaf rate*

In the previous section allusions were made to the role of LAI in determining the overall yield of the crop. Straightforward reasoning suggests that this overall yield is controlled both by the efficiency of the leaves of the crop as producers of dry material and by the leafiness of the crop itself. Both of these concepts have been dealt with individually, and their important joint role will be considered in the following section.

4.4 Crop growth rate

4.4.1 Theoretical aspects

In section 3.3.1 instantaneous unit leaf rate, **E**, was defined as

$$\mathbf{E} = \frac{1}{L_A} \cdot \frac{dW}{dT} \tag{3/8}$$

and in section 4.3.1 we saw that leaf area index, **L**, (also an instantaneous value), was defined as

$$\mathbf{L} = \frac{L_A}{P}. \tag{4/1}$$

Multiplying these two together we get

$$\frac{1}{P} \cdot \frac{dW}{dT} = \frac{1}{L_A} \cdot \frac{dW}{dT} \times \frac{L_A}{P}. \tag{4/2}$$

The new quantity on the left-hand side of equation 4/2 is the instantaneous rate of dry matter production per unit area of land, a simple and important index of agricultural productivity. Although various workers had previously made use of this concept, it was first given a name, crop growth rate, by WATSON (1958). The contraction CGR is now in general use and **C** is a convenient symbol for its instantaneous value. Hence, instantaneously,

$$\text{CGR} = \text{ULR} \times \text{LAI} \tag{4/3}$$

or

$$\mathbf{C} = \mathbf{E} \times \mathbf{L}. \tag{4/4}$$

The units of CGR are weight land area^{-1} time^{-1}.

Equation 4/4 is the central relationship in the study of population and community growth in the same way that

$$\mathbf{R} = \mathbf{E} \times \mathbf{F} \tag{3/13}$$

is central to the study of the growth of plants as individuals. One cannot press the similarity between equations 4/4 and 3/13 too far since **C** is closer in concept to absolute growth rate than to **R**, and **L** and **F** may each be applied to the analysis of crop growth in their own right instead of being analogues as suggested here. Nevertheless, in general terms it is easy to see that in each case the overall performance of the system (**C** and

R respectively) is broken down into two components: the productive efficiency of its leaves (E in both cases) and its leafiness (L and F respectively).

4.4.2   Practical approaches

As in the cases of R, E and F, the two notions of instantaneous value at a given time and mean value of over a given time interval need to be borne clearly in mind. Perhaps the most satisfactory way of deriving the instantaneous CGR, C, is to multiply together the instantaneous values E and L, using the relationship defined in equation 4/2. The derivation of E is discussed in the next chapter (section 5.4) but the instantaneous value L is available directly from raw data (equation 4/1).

Dealing with mean values, we already have the relationship

$$_{1-2}\bar{E} = \frac{_2W - {}_1W}{_2T - {}_1T} \cdot \frac{\log_e {}_2L_A - \log_e {}_1L_A}{_2L_A - {}_1L_A} \qquad (3/9)$$

and a mean value for L between times 1 and 2, $_{1-2}\bar{L}$, may be estimated as

$$_{1-2}\bar{L} = \frac{_1L + {}_2L}{2} \qquad (4/5)$$

if linearity between L and time is assumed (cf. the treatment of F in equation 3/11).

A mean crop growth rate, $\bar{C}$, may be calculated, without recourse to ULR and LAI, as

$$_{1-2}\bar{C} = \frac{1}{P} \cdot \frac{_2W - {}_1W}{_2T - {}_1T} \qquad (4/6)$$

where $_1W$ and $_2W$ are the dry weights of crop harvested from equal (but separate) areas of ground, P, at times 1 and 2. If $_1W$ and $_2W$ are each expressed per unit quantity of P then equation 4/6 can be simplified to

$$_{1-2}\bar{C} = \frac{_2W - {}_1W}{_2T - {}_1T}. \qquad (4/7)$$

Mean crop growth rate in this form indeed becomes an absolute growth rate – a difference in size divided by a difference in time (see section 3.1 and equation 3/7).

These calculations for the mean values $\bar{C}$, $\bar{E}$ and $\bar{L}$ all depend on various assumptions about the relationship between the raw variables and time (see, for example, section 3.3.3). Only rarely do these different assumptions concur, so, in most cases,

$$\bar{C} \neq \bar{E} \times \bar{L} \qquad (4/8)$$

for just the same type of reason that

$$\bar{R} \neq \bar{E} \times \bar{F}. \qquad (3/15)$$

The mean values of all of these quantities are nevertheless extremely useful, and can be derived simply. Even if their interrelationship is seldom exact it is often close enough to be valuable in interpreting the overall growth of the system in terms of its component processes. Statistical analyses may be performed on collections of mean values each derived from 'paired' raw data, as before (section 3.2.3).

4.4.3   Crop growth rate, unit leaf rate and leaf area index in operation

As we have seen in section 3.3.3 ULR normally declines in magnitude as a crop approaches maturity and LAI (section 4.3.2) normally increases. Indeed, the former trend is, at least in part, a consequence of the latter. Now, since CGR is the product of ULR and LAI, the direction and extent of its own drift with time depends on the relative magnitude of these trends. STOY (1965) grew *Triticum aestivum* (wheat) under controlled-environment conditions and was able to show that as plants grew in the vegetative phase, increasing $\bar{L}$ caused less than proportionate decreases in $\bar{E}_A$ (Fig 4-3), with the result that $\bar{C}$ increased with time.

BLACKMAN (1968), again quoting unpublished results of J. Newton on the growth of *Gladiolus*, demonstrated that increased LAI at high densities of planting led to increased CGR despite a fall-off in ULR. Conversely, WATSON (1958) removed different fractions of the foliage of *Brassica* (kale) and *Beta vulgaris* (sugar beet) crops growing at Rothamsted and found that increased values for ULR resulted. CGR itself was maximal at the intermediate values of LAI in kale (3 to 4) but in sugar beet Watson reported that maximal CGR occurred above the highest value of LAI encountered in the experiments (5.8). Optimal LAI was probably in the range 6 to 9. Judicious thinning of kale crops, but not sugar beet, might thus increase total dry matter yield.

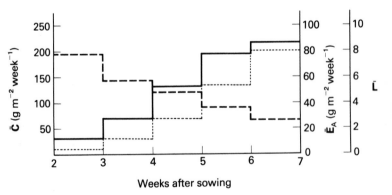

Fig. 4–3   Mean values of crop growth rate, $\bar{C}$ (——); unit leaf rate, $\bar{E}_A$ (– – –); and leaf area index, $\bar{L}$ (· · · · ·) in relation to age, in wheat. (Plotted from data given by STOY, 1965.)

WATSON (1952) discussed the relative importance of variation in LAI and ULR in determining CGR and concluded that LAI is on the whole more important because it is more strongly dependent on the environmental conditions and management regime of the crop.

Inter-varietal differences are well known, for example WATSON (1958), reworking data of WATSON (1947), showed that $\bar{E}$ and $\bar{L}$ were inversely related in a selection of five varieties of *Solanum tuberosum* (potato). In this particular example, $\bar{E}$ varied by a greater proportion than $\bar{L}$ and was thus the more important determinant of $\bar{C}$.

4.5   Leaf area duration

4.5.1   *Theory and practice*

When leaf area index is plotted against time (e.g. Fig. 4–1) the resulting curve allows not only an examination of the time-course of this quantity but also an estimate of what WATSON (1947) called the 'whole opportunity for assimilation' that the crop possesses during the period in question. Watson suggested that the integral of (the area lying beneath) the LAI v. time curve might conveniently be called leaf area duration, LAD, since it 'takes account both of the magnitude of leaf area and its persistence in time'. In effect, it represents the leafiness of the crop's growing period. The single symbol in use for LAD is **D**.

Three methods exist for estimating LAD from the LAI v. time curve:

(i) if the equation of the curve is known, or can be derived, then those with a knowledge of calculus can determine the true value of **D** numerically, by integrating the curve between the upper and lower time limits which encompass the period of interest (MACHIN, 1976, p. 84; CAUSTON, 1977, p. 139).

(ii) The curve may be traced on to opaque paper where it and the area beneath it may be cut out with scissors. The area of this resulting shape may then be determined by any of the methods suggested in section 2.2 for the determination of leaf area.

(iii) An approximate value for **D** may be derived graphically in the manner shown in Fig. 4–4. The area of the shaded trapezium, which approximates to the area under the curve between times $_1T$ and $_2T$, is its breadth times its average height, or

$$_{1-2}\mathbf{D} = \frac{(_1\mathbf{L} + _2\mathbf{L})(_2T - _1T)}{2}. \tag{4/9}$$

The area under the whole curve, or under that part of it which is of interest, may be estimated as the sum of all such trapezia, provided that these are constructed in such a way that their top edges approximate closely to the shape of the curve itself. That is, where the curvature is marked the trapezia should be narrow.

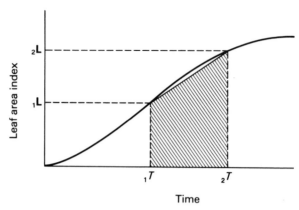

Fig. 4–4 Leaf area duration (shaded area) determined from a plot of leaf area index against time (see equation 4/9.)

The units of any area are the product of the units of each of its two dimensions. Since LAI is a dimensionless ratio, the units of LAD are the same as those of time. For example, if the area beneath a LAI v. time curve indicates a LAD of 15 weeks this might result from foliage equivalent in area to the whole ground area covered by the crop being sustained for a period of 15 weeks. Or from foliage of area greater on average than this ground area being sustained for a proportionately shorter number of weeks (multiplying up to the same total LAD), and so on.

If the LAD of a crop and its mean ULR are known then its final yield may be predicted. Less perversely, if this yield is already known (as it would be if ULR had been derived) then it may be broken down into these two components:

$$\underset{\text{(weight area}^{-1})}{\text{Yield}} \simeq \underset{\text{(time)}}{\text{LAD}} \times \underset{\text{(weight area}^{-1}\text{ time}^{-1})}{\text{ULR.}} \qquad (4/10)$$

The approximation sign is used here because the concept of mean unit leaf rate over a whole season is inevitably very crude (see section 3.3.3) and because, in the concept of leaf area duration itself, equal areas beneath the LAI curve are treated as being equally useful 'opportunities for assimilation' – an even more crude assumption in view of the changes that take place, both ontogenetically within the crop and climatically in its environment, during the course of the crop's growth. EVANS (1972, p. 224) discusses these problems further.

4.5.2 An alternative approach

Another way to derive LAD is to estimate the area under the curve of L_A (as opposed to LAI) v. time. This produces another LAD, with units of area × time. For example, these might be m^2 week ('square metre weeks').

In effect, this is a LAD expressed in terms of leaf area per plant, instead of the leaf area per plot ('ground area equivalents') used in the previous section. This change in units is reflected in the general relationship

$$\text{Yield} \simeq \text{LAD (leaf area basis)} \times \text{ULR.} \qquad (4/11)$$
$$\text{(weight)} \quad \text{(area} \times \text{time)} \quad \text{(weight area}^{-1}\text{ time}^{-1})$$

Yield, like LAD, is now expressed on a per crop basis. The same inexactness applies here as applies to equation 4/10.

4.5.3 Analogous concepts

A number of concepts analogous to LAD may be envisaged. For example, KVĚT, SVOBODA and FIALA (1969) derived a biomass duration, BMD, to which KVĚT, ONDOK, NEČAS and JARVIS (1971) gave the symbol **Z**. This was defined as the area under the weight v. time curve in the same way that LAD was defined as the area under the same curve for LAI (or L_A). The units of BMD are weight × time. For example, these might be g week ('gram weeks'). This quantity is related to total yield, not by mean ULR, the productive efficiency of unit amounts of leaf area, but by the mean RGR itself, the productive efficiency of unit amounts of dry matter:

$$\text{Yield} \simeq \quad \text{BMD} \quad \times \quad \text{RGR} \qquad (4/12)$$
$$\text{(weight)} \quad \text{(weight} \times \text{time)} \quad ([\text{weight weight}^{-1}]\text{ time}^{-1})$$

This relationship is also inexact (cf. equations 4/10 and 4/11) but as a crude summary of the behaviour of the crop it can be useful. Květ, Svoboda and Fiala described BMD as 'being an approximate measure of the stand's vitality' and pointed out that

$$\text{LAD (leaf area basis)} \simeq \quad \text{BMD} \quad \times \quad \text{LAR.} \qquad (4/13)$$
$$\text{(area} \times \text{time)} \quad \text{(weight} \times \text{time)} \quad \text{(area weight}^{-1})$$

Parallel developments, using length, volume, total chlorophyll or calorific content as the raw data, will be obvious.

4.5.4 Leaf area duration in operation

Using the same four species grown at Rothamsted, WATSON (1947) demonstrated that LAD was a more important factor in determining final yield than mean ULR (Table 8). Equation 4/10, with yield and LAD as known quantities, was used to derive mean ULR. The analyses for sugar beet run only until the end of October, which is rather short of its whole growing season.

The time of year at which most of the crop's foliage is displayed is of some importance. Obviously, other things being equal the greatest possibilities for high production occur when a substantial LAI coincides with the midsummer conditions, where ULR is maximal (cf. Figs. 3–5 and 4–1). In Watson's group of four species this conjunction was seen only in barley.

Table 8 A comparison of yield, leaf area duration and unit leaf rate in four crops grown at Rothamsted. (From WATSON, 1947.)

	Yield tonne ha^{-1}	LAD weeks	Mean ULR tonne ha^{-2} week^{-1}
Hordeum vulgare (barley)	7.3	17	0.43
Solanum tuberosum (potato)	7.7	21	0.36
Triticum sp. (wheat)	9.5	25	0.38
Beta vulgaris (sugar beet)	12.0	33	0.36

4.6 Primary production, growth analysis and photosynthetic production

The concepts outlined in this chapter allow workers to describe and interpret certain aspects of the growth of plant populations and communities in a fairly precise way given only the simplest of raw data. But these techniques form only part of a wider spectrum of scientific activity in which the general aim is throughout to relate plant performance to environment.

On the broadest scale this study involves an assessment of the primary production of vegetation in the field, considered at the ecosystem level of organization. An outline of the techniques involved here has been given for herbaceous communities by MILNER and HUGHES (1968) and SINGH, LAUENROTH and STEINHORST (1975). The role of primary production in the energetics of the whole ecosystem has been summarized by PHILLIPSON (1966). In comparison with studies at this level, the growth analysis of populations and communities focuses attention on much less extensive processes, both in time and in space. Its emphasis is on the more specific and on the more detailed. It has both the disadvantage of providing only a limited, short-term view of events and the advantage of enabling a more precise idea of the nature of the plant/environment interaction to be gained.

In the other direction, studies of plant photosynthetic production – the 'applied' aspects of photosynthesis research – aim often to investigate the plant/environment relationship at the level of the leaf, leaf segment and chloroplast. The various methods appropriate for studies here have been reviewed by ŠESTAK, ČATSKY and JARVIS (1971). In comparison, plant growth analysis suffers the disadvantage of providing little information about the fundamental physiological processes that govern the reactions of plants to environmental factors, even though valuable clues, if not detailed explanations, may sometimes emerge. On the other hand, the great advantage of many of the quantities involved in plant growth analysis is that they provide accurate measurements of the sum performance of the plant integrated both throughout the whole plant and across substantial intervals of time. To predict this from the starting point of purely physiological observations would involve many dangerous assumptions. In plant growth analysis the population or community is judged more by results than by promises. For example, if a deduction from W is made to allow for the non-carbon (mineral ash) content of the plant, the mean unit leaf rate is a true estimate of the net photosynthetic activity of the crop over any required time interval.

In reality these three approaches form a continuum. The investigator enters this continuum at a level appropriate both to the aims of his investigation and to the facilities that he has at hand.

4.7   Demography and population dynamics

As a postscript to this brief review of the growth analysis of populations and communities mention should be made of the complementary field of demography, on which the study of population dynamics is built. Developed mainly around studies on the human population its aim is more to describe and interpret the changes that occur in numbers of individuals rather than their changes in biomass per individual or their rates of functioning expressed on a unit basis. KEYFITZ (1968) has written a mathematical introduction to this field. In biology, animal science has borrowed more heavily from demography, but there are important implications also for plant science in many of the concepts such as birth and death rates, survivorship and life expectancy (HARPER and WHITE, 1974). SOLOMON (1976) gives an excellent introduction to the biological applications of demography and SARUKHÁN and HARPER (1973) provide a recent botanical example of a study involving some of its techniques.

In plant science the two fields of demography and growth analysis were distinct until BAZZAZ and HARPER (1977) published an account of the growth of *Linum usitatissimum* (cultivated flax) in relation to shading and density of planting. These authors claimed that the application of life-table and other demographic analyses to *leaf* birth and death (in contrast to the more usual case where attention is paid to the birth and death of whole individuals) permitted a more sophisticated interpretation of plant responses to environmental factors than was possible through the medium of conventional plant growth analysis. Commenting on this conclusion, HUNT (1978) suggested that the undisputed advantages of the demographic approach, in this case, were due more to the increased amount of information gained when working below the level of the whole organism. Hunt proposed that a growth-analytical investigation of the rate of functioning of individual leaves, taken separately, might yield a still more detailed interpretation of events, if this was to be the objective. Demography and plant growth analysis were different in kind not in degree, suggested Hunt, and such comparisons of techniques as were possible could only be made at the same level of organization.

5 Computerized Growth Analysis

5.1 Use of computers in conventional analyses

Each of the growth analysis quantities so far discussed can be calculated as a mean value across a harvest interval by the direct substitution of raw data into a formula. Occasionally, instantaneous values may also be calculated. If only a few values of such quantities are required from a body of experimental data the computational labour will not be great. But as the complexity of the experimental study increases then a more computerized approach becomes essential.

Technological developments in the field of desk-top or portable computers or calculators have moved swiftly and it is now possible to buy sufficient computing power to support small experimental programmes very cheaply indeed. For example, only the most basic four-function calculator is needed to derive mean absolute growth rate (section 3.1 and equations 3/7, 4/7), mean or instantaneous leaf area ratio (equations 3/10, 3/11), specific leaf area and leaf weight ratio (section 3.4.1), root/shoot ratio, root (or shoot) weight fraction and fresh weight/dry weight ratio (section 3.5), and leaf area index (equations 4/1, 4/5). If a single memory register is available then the following may be derived without the need to write down intermediate stages in the calculation: mean relative growth rate, using natural logarithms supplied either from tables or from the calculator, (equation 3/5) and mean leaf area duration (equation 4/9). If more than one memory is available then mean unit leaf rate (equations 3/9 and 3/19) and mean values of the analogous quantities discussed in section 6.2 can be derived in one operation.

Some of the more sophisticated small calculators are programmable (a series of instructions telling the calculator what to do may be inserted to last throughout the whole calculating session). This is a great advantage in handling the more complex formulae, such as those referred to above for mean unit leaf rate. In some cases programmable calculators are also able to handle simple regression analyses, bringing into convenient reach the allometric relationships described in section 3.6.

If the experimental scheme is lengthy or elaborate, but at the same time it is not desired to adopt the fully-computerized approach discussed in the remainder of this chapter, then it may be worth developing a more comprehensive approach to the handling of growth-analytical data in the conventional way. For example, a calculator program may be devised in which several variables, such as the individual weights of different parts of a harvested plant and its total leaf area, are inserted. The same data are then supplied for a 'paired' plant from another harvest (section 3.2.3). The various rates and ratios may then be calculated and, given sufficiently advanced equipment, stored for statistical analysis later. To make this approach really worthwhile, data in a machine-readable form, such as paper or magnetic tape, would be valuable and a permanent record of results on a tape or print-out would be all but essential. It is in the interests of the experimenter to devise analytical procedures which use the available computing hardware to its limit.

5.2 Curve fitting – the functional approach to growth analysis

5.2.1 Rationale

The calculation of the various growth analysis quantities as mean values over the period of time intervening between two harvests has been the standard approach for most of the sixty years that have elapsed since the early origins of the subject. The calculations are straightforward and if the results obtained are sometimes approximate, then this defect can be minimized by an experimental design which is sympathetic to this analytical approach. More recently, however, many workers have taken advantage of high-speed computing facilities to fit mathematical functions to experimental data. These describe the relationship between the data and, say, time. From these functions ('growth curves'), *fitted* values of data are extracted and then used to derive the various growth analysis quantities which may subsequently be plotted as fitted instantaneous values. The single great opportunity seized upon in this approach is that the central relationships

$$RGR = ULR \times LAR \qquad (3/12)$$

and

$$CGR = ULR \times LAI \qquad (4/3)$$

are employed without approximation. The instantaneous values of these quantities are, by definition, related in the way expressed in equations 3/12 and 4/3.

5.2.2 Advantages

If one abandons the commitment to calculations on the classical harvest interval basis (which normally involves relatively few harvests of relatively large size) then several advantages accrue if one is able to adopt what RADFORD (1967) has called the 'dynamic' approach to plant growth analysis:

(i) The various assumptions required to calculate mean values, for example of unit leaf rate (section 3.3.2) or of leaf area ratio (equation 3/11), are evaded completely.

(ii) More frequent harvests of fewer plants may be used to provide information about the growth of the plants on a more or less continuous basis (HUGHES and FREEMAN, 1967).

(iii) If (ii) is adopted then less information is at risk from accidental loss at any one harvest and the work-load is more evenly spread throughout the period of measurements.

(iv) The 'pairing' of plants across the harvest intervals becomes unnecessary. (In the case of complex experiments this avoids real problems. For example, in the procedure outlined in section 5.1 for the handling of many variables, with respect to which variable is the pairing to be done?)

(v) Small deviations from the overall trend of the original experimental data against time are 'smoothed' often making the final results less erratic (HUNT, 1973).

(vi) Effectively, all of the available information is used in obtaining any one value of the derived quantities, since these are estimated not from raw data, but from fitted values of the raw data supplied by the chosen growth functions.

(vii) Statistical analyses may be integrated into the same analytical procedure as the calculation of the derived quantities.

5.2.3   Assumptions and prerequisites

In purely growth-analytical terms the only assumption necessary for the adoption of this approach is that the fitted growth curves adequately describe the trends in the raw data. This, in turn, depends on the assumption that the raw data adequately describe what is really happening in the plants under investigation. In a statistical sense there are clear procedures laid down for determining whether a chosen curve is or is not a good functional model of the data in hand. PARKER (1978, § 9.4) provides an introduction to these.

5.3   A simple example of the functional approach

The foregoing section (5.2) introduces several new concepts and it is probably best, before going on to examine more complete examples, to deal in some detail with a simple case of the functional approach. This illustrates the relationships between the raw data, the growth function, the derived data and the statistical analyses.

One of the species grown in the experiments of GRIME and HUNT (1975) was the grass *Holcus lanatus* (Yorkshire fog). Favourable controlled-environment conditions were used and five plants each were harvested at 11, 20, 28 and 35 days after planting. Figure 5–1 provides a plot of the natural logarithms of the total dry weights of these plants against time.

To derive an instantaneous relative growth rate from these raw data requires that we fit a growth function to them. In other words, we need an expression of the general form

$$\log_e W = f(T) \tag{5/1}$$

since we saw in section 3.2.3 that the slope of such a function, $f'(T)$, would provide the instantaneous value of relative growth rate, **R**. Taking one of the simplest possible functions, a linear regression, we can derive an equation of the form

$$\log_e W = a + bT \tag{5/2}$$

from the raw data, where a and b are constants with values particular to this case. Using the data given in Fig. 5–1 we regress the twenty values of $\log_e W$ (mg) on to their corresponding values of T (zeroed to 11 days and converted to weeks) and find that

$$\log_e W = 2.34 + 1.56\,T \tag{5/3}$$

This regression line is shown in Fig. 5–1. Now, we have seen that the slope of this regression is the value required since

$$\mathbf{R} = \frac{d(\log_e W)}{dT}. \tag{3/3}$$

This slope is obtained by differentiating equation 5/2 (CAUSTON, 1977, p. 87):

$$\mathbf{R} = \frac{d(\log_e W)}{dT} = b \tag{5/4}$$

or, in the case in hand,

$$\mathbf{R} = \frac{d(\log_e W)}{dT} = 1.56. \tag{5/5}$$

Thus, at any time, $\mathbf{R} = 1.56$. The units are week^{-1} (derived from the units of T). In the case of more complex models, such as will be met in succeeding examples, T remains in the expression for **R**; in other words **R** is then time-dependent.

The model that we have chosen is a reasonable fit to the raw data even though there is a suspicion from Fig. 5–1 that a curve would be more appropriate. However, since its main purpose is to illustrate the functional technique itself and to enable comparisons with conventional analyses to be made, we will accept it for the moment: it is certainly not a bad fit.

Among the parameters that most regression programs provide is the standard error, SE, of the coefficient b. (If the regression has been calculated by hand then this SE may easily be derived from the intermediate quantities required to compute the regression itself.) This value is, of course, also the SE of **R** and for equation 5/3 it is equal to 0.046

week^{-1}. Multiplying by Student's t (at $P < 0.05$ and $n - 2 = 18$ d.f.) we get the 95% confidence limit, ± 0.096 week^{-1}. This delimits the range within which the value of **R** would lie on 95% of occasions in an infinite series of identical experiments. That is, 1.45 to 1.66 week^{-1}. In the present example this value holds throughout the period of measurement because we have fitted the log-linear (exponential) model (see section 3.2.3).

Table 9 compares this fitted value of **R** with mean values, $\bar{\mathbf{R}}$, for the various harvest intervals calculated by way of the conventional equation 3/5 using mean values of W at each harvest occasion.

Table 9   Mean and instantaneous relative growth rates in *Holcus lanatus* (Yorkshire fog). (Data of GRIME and HUNT, 1975.)

Quantity	Symbol (including time interval, days)	Value (week^{-1})
Mean relative growth rate	$_{11-20}\bar{\mathbf{R}}$	1.75
	$_{20-28}\bar{\mathbf{R}}$	1.59
	$_{28-35}\bar{\mathbf{R}}$	1.46
	$_{11-35}\bar{\mathbf{R}}$	1.61
Instantaneous relative growth rate	$\mathbf{R}$	1.56 throughout

The three mean values $_{11-20}\bar{\mathbf{R}}$, $_{20-28}\bar{\mathbf{R}}$ and $_{28-35}\bar{\mathbf{R}}$ decline slightly with time, reinforcing the previous suspicion that the plot of $\log_e W$ on T may not be entirely linear. The overall mean value, $_{11-35}\bar{\mathbf{R}}$, does not agree exactly with **R** itself since the former is based only on half of the experimental data (the 11-day and 35-day harvests) whereas the latter involves them all.

In cases where the plot of $\log_e W$ against time is clearly non-linear it may still be valid to fit the linear model if the experimenter needs a single value of **R** for comparative purposes. Obtained in this way **R** is in one sense a mean value, $\bar{\mathbf{R}}$: the average slope of the whole plot of $\log_e W$ on time. The difference between this and, say, the $_{11-35}\bar{\mathbf{R}}$ given above, almost warrants the introduction of a distinct name and symbol. The latter is the true mean slope between the two end points of the plot – quite distinct in concept from the average slope of the whole plot.

5.4   More complete approaches

5.4.1   *Introduction*

The foregoing example illustrates how, for the simplest possible case, an instantaneous value of **R** may be derived from a fitted growth function. The following sections trace the more important developments in this field (for a fuller bibliography see HUNT and PARSONS, 1974). They are concerned both with more complex growth functions and the derivation not only of **R** but also of **E** (instantaneous unit leaf rate) and of **F** (leaf area ratio). The computing support necessary to make use of these different approaches varies, according to the complexity of the procedure, from a small machine on which simple regressions can be performed in easy stages to a full-scale computing installation able to accept programs of several hundred statements written in high-level languages such as FORTRAN and ALGOL.

5.4.2   *VERNON and ALLISON (1963)*

Presenting data on the growth of *Zea mays* (maize), these authors fitted second order polynomials (quadratic curves) to W and to L_A:

$$W = a + bT + cT^2 \tag{5/6}$$

and

$$L_A = a' + b'T + c'T^2 \tag{5/7}$$

where a, a', b, b', c, c' are constants.

Differentiating equation 5/6 they obtained the absolute growth rate

$$\frac{dW}{dT} = b + 2cT \tag{5/8}$$

from which instantaneous values of unit leaf rate were derived in this manner

$$\mathbf{E} = \frac{1}{L_A} \cdot \frac{dW}{dT} = \frac{b + 2cT}{a' + b'T + c'T^2}. \tag{5/9}$$

Solving equation 5/9 for different values of T enabled Vernon and Allison to plot **E** v. T on a continuous basis throughout the period studied.

Although straightforward in its approach, and making an important step forward, this method suffers from the serious statistical drawback that progressions of W and L_A against time (as opposed to those of $\log_e W$ and $\log_e L_A$) seldom show a uniform variability with increasing T, which they must do if they are to be subjected to regression analysis. In statistical terminology such data are usually *heteroscedastic*: the variance of each sample is not constant from harvest to harvest. This being the case it would be wise to employ this approach, only where the reverse, *homoscedasticity*, can be demonstrated. This is likely to be only in very short-term series of measurements.

5.4.3   *HUGHES and FREEMAN (1967)*

These authors also made an important step forward when they employed polynomial regressions to fit curves to logged experimental data. Using data on the growth of *Callistephus chinensis* (Chinese aster) under controlled-environment conditions, they developed the following third-order (cubic) regressions:

$$\log_e W = a + bT + cT^2 + dT^3 \tag{5/10}$$

and

$$\log_e L_A = a' + b'T + c'T^2 + d'T^3. \tag{5/11}$$

Here Vernon and Allison's family of constants is joined by d and d'.

Because logarithms were employed, heteroscedasticity was not a problem and the third-order (cubic) polynomials enabled quite complex trends in the (logged) raw data to be followed. Hughes and Freeman derived **R**, **F** and **E** in this manner:

$$\mathbf{R} = \frac{d(\log_e W)}{dT} = b + 2cT + 3dT^2, \tag{5/12}$$

$$\mathbf{F} = \frac{L_A}{W} = \text{antilog}_e (\log_e L_A - \log_e W) \tag{5/13}$$

that is,

$$\mathbf{F} = \text{antilog}_e (5/11 - 5/10) \tag{5/14}$$

and, using equation 3/13,

$$\mathbf{E} = \frac{\mathbf{R}}{\mathbf{F}} = \frac{5/12}{5/14}, \tag{5/15}$$

where the numbers represent references to equations.

For the first time, reliable estimates of instantaneous unit leaf rate became available and Hughes and Freeman were able to show that increasing the ambient CO_2 concentration of *Callistephus* from 325 ppm through 600 ppm to 900 ppm increased **E** significantly, $P < 0.05$ (Fig. 5–2).

However, one disadvantage of Hughes and Freeman's approach was that their computer program fitted cubic polynomials to all logged data in every case, whether or not a real cubic relationship existed. This had the effect of unnecessarily increasing the SES of the derived quantities in cases where the quadratic (c or c') or cubic (d or d') parameters of regression were not significant. There is something of this problem evident in Fig. 5–2 where the final upswing in **E** is more a result of the analyticial methods employed than a genuine feature of the plant's behaviour. The technique does not lie (the limits are wide enough for one to be able to discount this upswing) but the truth is blanketed.

§ 5.4

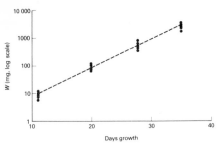

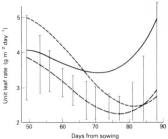

Fig. 5–1 The progression of whole plant dry weight on time in *Holcus lanatus*, grown in a productive, controlled environment. Four harvests, each of five replicates, are shown. The equation of the fitted line is equation 5/3. (Data of GRIME and HUNT, 1975.)

Fig. 5–2 Progress curves of instantaneous unit leaf rate in *Callistephus chinensis* (cv. Johannistag). Three levels of carbon dioxide concentration were used: 900 ppm, ——; 600 ppm, - - - - -; 325 ppm, ——. 95% confidence limits are added to the 325 ppm curve. (From HUGHES and FREEMAN, 1967.)

5.4.4 NICHOLLS and CALDER (1973)

These authors published a general discussion on the use of regression analyses for the study of plant growth. In worked examples on the growth of *Atriplex spongiosa* and *A. lindleyi* (Australian species of orache) they demonstrated that increasing the complexity of the regressions used to describe the changes with time in logged plant variables increases the SES of the derived growth analysis quantities. It may also lead to spurious values of these quantities themselves, stated Nicholls and Calder, who summarized their approach by saying that '. . . each set of data has to be considered on its own for the selection of the appropriate regression model. Over fitting is a real trap . . .'.

5.4.5 HUNT and PARSONS (1974)

Working independently, but along similar lines to Nicholls and Calder, Hunt and Parsons developed Hughes and Freeman's approach a stage further. A stepwise regression procedure was incorporated in which polynomials of the first to third order were fitted to the logarithms of W and L_A. Tests were made to determine in each case which of these fits was best and for any one set of data it was also possible to fit different orders of regression to $\log_e W$ and $\log_e L_A$. Hence, Hunt and Parsons's equations were selected from a range that included both those of Hughes and Freeman (equations 5/10, 5/11) and the simpler possibilities

$$\log_e W = a + bT + cT^2 \qquad (5/10a)$$

$$\log_e W = a + bT \qquad (5/10b)$$

and

$$\log_e L_A = a' + b'T + c'T^2 \tag{5/11a}$$

$$\log_e L_A = a' + b'T. \tag{5/11b}$$

An additional facility, giving merely a mean and standard error for $\log_e W$ or $\log_e L_A$, was included for cases where no significant trends with time existed. From the best equation for $\log_e W$ and the best for $\log_e L_A$ ($P < 0.05$) the derived quantities **R**, **E** and **F** were calculated as in the Hughes and Freeman method.

A disadvantage of Hunt and Parsons's approach is that difficulties sometimes occur when trying to interpret a whole body of data in which regressions of varying order have been applied objectively, according to the needs of individual sets of data (HURD, 1977). Because the form of the progressions of the derived quantities on time depends immutably on the nature of the original regressions, and because small differences among the latter are often reflected in large differences among the former, comparisons between species or treatments are sometimes puzzling when quite distinct patterns in the behaviour of, say, **R** or **E** emerge from sets of raw data which differ only slightly from one another, but nevertheless enough to cause the selection of different growth functions. This problem is discussed further in section 6.4.3.

5.4.6   Other approaches

Although not wholly one of functional approaches to plant growth analysis, the methods of GOODALL (1949) should be mentioned. Seedlings of *Theobroma cacao* (cacao) were grown in an outdoor nursery at Tafo, Ghana. Polynomials were fitted to W v. T, not for the direct calculation of **R**, but to redescribe the experimental data in the form of a smooth progression from which values of W could be read off and inserted into the conventional harvest interval formulae. This procedure possesses some of the advantages of the functional approach (smoothing of data, possibility of interpolation) but not all of them (see section 5.2.2).

RICHARDS (1959) put forward a new mathematical expression, more complex than those we have considered so far, as a general-purpose model for fitting growth curves (from which, of course, other quantities may be derived). CAUSTON (1969) has devised a computer program for fitting this function which, when rewritten in the notation used in this booklet, appears as

$$W = a(1 + be^{-kT})^{-1/n}. \tag{5/16}$$

W is described as a function of T where a, b, k and n are constant for the case in hand and e is the base of natural logarithms. The function is a generalization of the logistic equation (see SOLOMON, 1976, p. 17) and it is possible that, of any single growth function, it may be the most widely useful, although whether or not this is the case has still to be established clearly. If a crude analogy were needed to assess this single flexible function in relation to the 'package' offered by Hunt and Parsons it might take the form of two questions: When confronted with nuts of all conceivable shapes and sizes, which is best, a single fully-adjustable spanner or a set of various fixed (or only slightly-adjustable) spanners? The adjustable spanner is certainly useful, especially in borderline cases, but can it beat the simple fixed spanner at its own game?

5.5   General applicability of the computerized approach

There is no doubt that the adoption of one or more of the computerized approaches is, for those with the inclination and facilities to set up the necessary analytical procedures, the best method for tackling data from large programmes of growth studies. Approaches such as that of Hunt and Parsons obviously require a full-scale computing installation if they are to be used often; 'For those less well endowed the work involved can easily become impossibly tedious' (EVANS, 1972, p. 343). However, the more accessible regions of the functional approach may easily be explored with the aid of a large programmable calculator or a desk-top computer. The potential benefits (section 5.2.2) are worth a careful assessment of local computing possibilities, however modest, while for those planning to work with high-level languages, NELDER (1975) has written a general guidebook with which beginners in this, or any other, field of computing will find most valuable.

5.6   Biological relevance of growth functions

The various growth functions that have been discussed so far have all been considered from one point of view only: are they good models of the raw data? The parameters of the equations have had no biological relevance, they were merely tools with which to construct the desired product, the derived quantities like **R**, **E** and **F**. They were *empirical* models and simply redescribed the data without giving rise to any information that was not contained in the data (THORNLEY, 1976).

Another approach to the modelling of plant growth may be termed the *mechanistic* approach. This is conceived in terms of the mechanism of the system, or how the parts of the system work together as they might in a machine. An extensive review of the process of dry weight increase from a mechanistic viewpoint has been provided by RICHARDS (1969). One such mechanistic model is the 'mono-molecular' function (borrowed from physical chemistry):

$$_TW = W(1 - be^{-kT}) \qquad\qquad (5/17)$$

where W is the final, limiting dry weight of the plant and $_TW$ its dry weight at time T. The absolute growth rate is $k(W - _TW)$. This is proportional to the amount of growth *yet to be made* and it declines linearly in value with increases in $_TW$.

Not surprisingly, mechanistic models seldom fit experimental data as well as empirical ones, firstly, because whole plants rarely behave as simple machines (except perhaps over short periods of time) and, secondly, because empirical models are designed expressly (and only) with accuracy of fit in view. The relationship between empiricism and mechanism in the whole field of plant modelling has been summarized by THORNLEY (1976): 'It needs to be stressed that there is no clearly defined dividing line between the two methods, and it is usual for most modelling exercises to contain both empiricism and mechanism in varying admixtures. It is more a matter of emphasis. The mechanistic modeller will tend to construct his models before doing the experiments, thinking of possible mechanisms and deducing their consequences by means of a model; the experiment will then test his hypotheses, and possibly favour one mechanism rather than another. . . . On the other hand, the empirical modeller may well make his guesses about mechanism after doing the experiment and looking at the data, so he begins an investigation as an empiricist and ends up as a mechanist. In practice the modeller swings like a pendulum between mechanism and empiricism; he tries to make progress whenever and however he can.'

In the context of the plant growth analysis described in this booklet, pure empiricism is sustained right through the functional approach as far as the calculation of the derived quantities (which are themselves only a special redescription of the original data). Only then, if these quantities form components of further model-building programmes relating, to give a recent example, nutrient uptake and growth response to independently measured soil and plant characteristics (NYE, BREWSTER and BHAT, 1975), would mechanism come into its own.

6 Developments in Plant Growth Analysis

6.1 The generalized approach

The foregoing concepts, with the exception of absolute growth rate and the allometric constant, each take one of three general forms:

(i) $(1/Y)(dY/dX)$, e.g. the various relative growth rates;
(ii) Z/Y, e.g. leaf area ratio, root/shoot ratio;
(iii) $(1/Z)(dY/dX)$, e.g. unit leaf rate, crop growth rate;

where Y and Z are general symbols which may stand for any of the raw data (the various weights and areas). X is also a general symbol representing an independent variable, such as time, against which these concepts are to be examined. If the experimenter has devised a general analytical scheme for the derivation of these he may, by controlling the identity of the raw variables input to this scheme, calculate a whole range of useful quantities using the one technique. HUNT and BURNETT (1973) provide an example of the development of a reasonably full series of derived quantities, using the methods of HUGHES and FREEMAN (1967).

6.2 Other quantities of the form $(1/Z)(dY/dX)$

6.2.1 Introduction

In simple words, these are 'rates of production of something per unit of something else'. In plant growth analysis, provided that the 'something', Y, is of interest to the experimenter and that the 'something else', Z, may reasonably be held responsible for its production, then $(1/Z)(dY/dX)$ is a tool of fundamental importance.

The following sections describe the use of this tool in a selection of guises. The list is not exhaustive, nor by any means have all of the possibilities in this direction been followed up.

6.2.2 A simple index – unit shoot rate

When discussing unit leaf rate, E, in section 3.3.3 we saw that it could be calculated on the basis of leaf weight (E_W), leaf area (E_A) or leaf protein content (E_P). This comprised an increasingly informative series for the experimenter interested in the detailed functioning of the plant, but at the cost of increased experimental labour.

HUNT and BURNETT (1973) devised a unit shoot rate (USR), the rate of production of dry weight per unit of shoot material. This may be notated as B and defined as an instantaneous value:

$$\mathbf{B} = \frac{1}{S_W} \cdot \frac{dW}{dT} \tag{6/1}$$

where, as before, S_W is the total dry weight of the above ground parts of the plant and W is the dry weight of the whole plant. Unit leaf rate may be regarded as a component of unit shoot rate since

$$\frac{1}{S_W} \cdot \frac{dW}{dT} = \frac{1}{L_A} \cdot \frac{dW}{dT} \times \frac{L_A}{S_W} \tag{6/2}$$

or

$$\mathbf{B} = \mathbf{E} \times \frac{L_A}{S_W}. \tag{6/3}$$

B, even more than $\mathbf{E_W}$, in relation to the aforementioned versions of **E**, provides a crude but simple index of the performance of the productive parts of the plant. The series **B**, $\mathbf{E_W}$, $\mathbf{E_A}$, $\mathbf{E_P}$ is thus one of increasing sophistication but decreasing ease of derivation. **B** is worth considering in two cases: (i) where it is difficult or unduly laborious to record L_W separately from S_W, and (ii) where there is little point in making this distinction (e.g. in young monocotyledons where leaves constitute a huge proportion of the above ground parts of the plant).

The values of **E** and **B** given for *Lolium perenne* (perennial ryegrass) by HUNT and BURNETT (1973) are correlated at $P < 0.001$ ($r = 0.907$, $n = 54$) despite the fact that they encompassed a wide range of experimental treatments, some of which included variation in L_A/S_W (equation 6/3). In this study, the response of **B** to the experimental treatments showed the same pattern as that of **E** although the amplitude of this response was reduced.

6.2.3   *Mineral nutrient uptake – specific absorption rate*

WILLIAMS (1948) commented that, in detailed studies on the mineral nutrition of plants, simple calculations yielding rates of uptake of mineral nutrients per day were complicated by what he called the 'size factor of the absorbing system'. He suggested that comparisons might be made on an equable basis if a concept analogous to unit leaf rate were to be employed. Instantaneously, he defined

$$\text{Rate of intake} = \frac{1}{R_W} \cdot \frac{dM}{dT} \tag{6/4}$$

where M is the plant's content of the mineral nutrient under consideration and R_W is the dry weight of the root system. The whole quantity provides an estimate of the rate of nutrient uptake per unit weight of root, a sort of root efficiency. For its calculation as an approximate mean value across a harvest interval, Williams used his 1946 formula (equation 3/9), with the necessary changes. WELBANK (1962) suggested the name specific absorption rate (SAR) for Williams's 'rate of intake', and used the symbol **A**. Hence, over the time interval $_1T$ to $_2T$:

$$_{1-2}\bar{A} = \frac{_2M - _1M}{_2T - _1T} \cdot \frac{\log_e {_2R_W} - \log_e {_1R_W}}{_2R_W - _1R_W}. \tag{6/5}$$

Like its counterpart, equation 3/9, this formula assumes a linear relationship between, in this case, R_W and M over the time interval in question.

The concept of specific absorption rate on a root weight basis is informative only for so long as R_W 'may reasonably be held responsible' for the intake of M. Perhaps root length, area, volume or number would be better? HACKETT (1969) and EVANS (1972, p. 228) discuss these problems further.

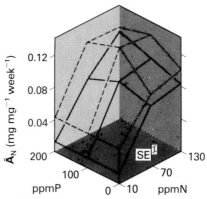

Fig. 6–1   Mean specific absorption rates for nitrogen in sugar beet grown in soil at three levels of added nitrogen and three of added phosphorus. Plants grown alone, – – – –; plants grown with competition from *Agropyron repens*, ———. *(From* WELBANK, 1964.)

An example of specific absorption rate in operation may be drawn from the work of WELBANK (1964) (Fig. 6–1). *Beta vulgaris* (sugar beet) was grown in soil in pots outdoors at Rothamsted. Three levels of added nitrogen and three of phosphorus were combined factorially and plants were grown both with and without competition from *Agropyron repens* (couch grass). At all levels of added nutrients, competition from *Agropyron* depressed the crop's mean specific absorption rate for nitrogen substantially ($\bar{A}_N$, root dry weight basis).

6.2.4   *Almost the converse – rate of mineral nutrient utilization*

Using arguments analogous to those of WILLIAMS (1948), KEAY, BIDDISCOMBE and OZANNE (1970) suggested that analyses of the utilization of mineral nutrients in plants, sometimes expressed as the reciprocal of the concentration of nutrients in the dry matter, would better describe the pattern with time in this utilization if expressed as the rate of dry weight increment per unit of absorbed nutrient. Unfortunately, Keay, Biddiscombe and Ozanne also used the symbol A for this quantity, conflicting with the **A** of WELBANK (1962). Perhaps it might be termed 'specific utilization rate' (SUR), and given the symbol U, so that, instantaneously

$$U = \frac{1}{M} \cdot \frac{dW}{dT} \tag{6/6}$$

and over the interval $_1T$ to $_2T$,

$$_{1-2}\bar{U} = \frac{_2W - _1W}{_2T - _1T} \cdot \frac{\log_e {_2M} - \log_e {_1M}}{_2M - _1M}. \tag{6/7}$$

Specific utilization rate bears this instantaneous relationship to specific absorption rate:

$$\frac{1}{R_W} \cdot \frac{dM}{dT} \times \frac{1}{M} \cdot \frac{dW}{dT} = \frac{1}{M} \cdot \frac{dM}{dT} \times \frac{1}{W} \cdot \frac{dW}{dT} \times \frac{W}{R_W} \tag{6/8}$$

or

$$\mathbf{A} \times \mathbf{U} = \mathbf{R_M} \times \mathbf{R_W} \times \frac{W}{R_W} \tag{6/9}$$

where $\mathbf{R_M}$ and $\mathbf{R_W}$ are the relative growth rates in mineral nutrient content and in whole plant dry weight respectively. W/R_W is, of course, the reciprocal of the root weight fraction (section 3.5). As constructed, specific utilization rate is a measure of the efficiency with which dry weight is increased by mineral uptake, although clearly this concept needs to be handled with care since the causal connection between the two processes is not exclusive.

6.2.5   *An example from sub-cellular biology*

In a study of the dynamics of leaf growth in *Trifolium subterraneum* (subterranean clover) WILLIAMS (1975, p. 179) developed a general purpose rate of production for one sub-cellular component of the leaf per unit of another component. This he termed **G**, and appended appropriate subscripts to describe the particular rate under examination. These subscripts serve to distinguish it from absolute growth rate, **G**. One example of such a rate is

$$\mathbf{G}_{PN, RNA} = \frac{1}{RNA} \cdot \frac{dPN}{dT}. \tag{6/10}$$

This represents the instantaneous rate of production of protein nitrogen, *PN*, per unit weight of ribonucleic acid, *RNA*. Over the time interval $_1T$ to $_2T$:

$$_{1-2}\bar{G}_{PN, RNA} = \frac{_2PN - _1PN}{_2T - _1T} \cdot \frac{\log_e {}_2RNA - \log_e {}_1RNA}{_2RNA - _1RNA}. \qquad (6/11)$$

Williams was able to show that in young clover leaves this functional efficiency of *RNA* as a producer of *PN* declined markedly with time – a valuable addition to the concept of ontogenetic drift (Table 10).

Table 10   Mean rates of production of protein nitrogen per unit of ribonucleic acid, $\bar{G}_{PN, RNA}$, in leaves of *Trifolium subterraneum*. (Data of WILLIAMS, 1975.)

Time interval (days)	$\bar{G}_{PN, RNA} (day^{-1})$	Mean values for group
10–11	5.8 ⎫	
11–12	5.7 ⎬	5.6
12–13	5.3 ⎭	
13–14	4.1 ⎫	
14–15	4.6 ⎪	
15–16	4.7 ⎬	4.4
16–17	4.4 ⎪	
17–19	4.4 ⎭	
19–22	1.8 ⎫	1.5
22–25	1.1 ⎭	

Williams surmised that since values of $\bar{G}_{PN, RNA}$ fell roughly 'into three groups, with mean values of 5.6, 4.4 and 1.5 day^{-1} for days 10–13, 13–19 and 19–25 respectively', this indicated that 'the synthesis of specific proteins or groups of proteins might be dominant for successive stages of development.'

6.3   Other independent variables

Where, for any one species, derived quantities are plotted against time (e.g. Fig. 3–4), the effects of any experimental treatments may accelerate or decelerate the natural ontogenetic drift in the quantity under examination. Hence plants of an age but in different treatments are not necessary all in the same morphogenetic condition. To overcome this disadvantage (minimizing the effects of ontogenetic drift) it may sometimes be useful to plot the derived quantities against another index of development, such as total dry weight, instead of against time (e.g. Fig. 3–3).

In practice this may be achieved if values of the derived quantities, like **R** and **E**, are plotted in order of increasing *W*. Naturally, this approach is more easily executed using instantaneous values. Indeed, if *W* does not figure in the derived quantity itself (e.g. sar, $(1/R_W) (dM/dT)$) it may actually be inserted into the analyses in place of T (giving $(1/R_W)$ (dM/dW)). However, these arguments are as broad as they are long: what guarantee is there that plants of one dry weight in different treatments have reached that dry weight by similar routes and are currently in the same morphogenetic condition?

There is also another danger here. When quantities derived, say, from W are plotted against W or against other quantities derived from it, misleading relationships can be obtained. If, for example, **R** is roughly constant, plotting **E** against **F** (that is, against **R/E**) can produce a straightish line for mathematical reasons alone. This can obscure a quite genuine dependence of **E** on **F** when increases in **F** lead to self-shading, thus causing a fall in **E**.

Another interesting development in this field considers growth not in relation to time, but in relation to the available photosynthetically-active radiation (EMECZ, 1962).

6.4 Problems with the computerized approach

6.4.1 Variability of data

ELIAS and CAUSTON (1976) discussed the question of which degree of polynomial should be fitted to logarithmically transformed raw data, when employing the functional approach to plant growth analysis (section 5.2). By artificially manipulating data on the growth of several species these authors showed that when growth curves were fitted to data of low variability, unrealistically high degrees of polynomial were quite likely to be selected if purely objective criteria alone were used. When variability was high, however, the 'best fit' was more likely to be a lower-order polynomial. Elias and Causton concluded that regressions should be fitted using harvest mean values where 'the test of adequacy of fit is independent of the underlying population variability.'

6.4.2 Analyses in segments

Given that the low-order polynomials are more satisfactory from many points of view than those of higher order (HUNT and PARSONS, 1974), problems arise when the experimenter is faced with extensive sets of data which may quite probably justify a high-order fit. For example, when the data on the growth of *Zea mays* (maize) in Fig. 1–1 were analysed by HUNT and PARSONS (1977) using HUNT and PARSONS's (1974) methods, a cubic polynomial was selected as being the best to fit to $\log_e \overline{W}$. However, as Fig. 6–2a shows, there were serious systematic deviations between the fitted regression and the original data.

Instantaneous relative growth rates, **R**, derived from Fig. 6–2a are plotted in Fig. 6–2b. The progression of **R** on time followed the progression of **R̄** (Fig. 3–2) inasmuch as a period of high **R** intervened between periods of negative **R**. But several points of detail present in Fig.

3–2 were smothered in the treatment given in Fig. 6–2b. For example, (i) the phase of negative **R** should be most marked at the beginning of the growth period, not at the end (Fig. 3–2); and (ii) because of this, the whole distribution of **R** should be skewed towards the early stages of growth and not almost symmetrical, as in Fig. 6–2b.

HUNT and PARSONS (1977) broke these data into two segments for separate analysis, 0–36 days and 36–113 days inclusive. Log.-cubic and log.-quadratic fits respectively were obtained ($P < 0.05$). Figure 6–2c shows that these regressions fitted the raw data in all important respects. The estimates of **R** derived from Fig. 6–2c, given in Fig. 6–2d, agreed closely with those given in Fig. 3–2 with the advantages of a smoothing of small deviations and the addition of errors (much smaller, incidentally, than those in Fig. 6–2b). The slight discrepancy between the two estimates

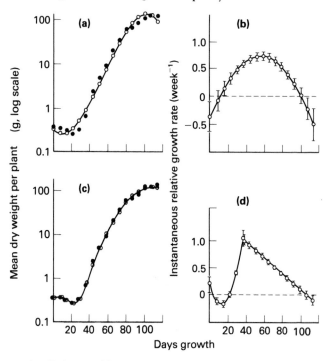

Fig. 6–2 'Badischer Früh' maize grown at Poppelsdorf in 1878: HUNT and PARSONS'S (1977) analysis using the methods of HUNT and PARSONS (1974). (a) Cubic regression fitted to the logarithms of $\overline{W}$ versus time; (b) instantaneous relative growth rates derived from (a); (c) cubic regression fitted to the period 0–36 days and quadratic regression fitted to the period 36–113 days inclusive; (d) instantaneous relative growth rates derived from (c). 95% limits are attached. Solid symbols: observed data; open symbols: fitted values. See also Figs. 1–1 and 3–2. (Data from KREUSLER, PREHN and HORNBERGER 1879.)

of **R** at 36 days was not significant at $P < 0.05$. Had this discrepancy been large, it would have implied that the juncture of the two regressions had been placed inappropriately.

In general, very lengthy or complex series of data could be approached by this 'segmentation' process provided only that in each segment there was enough information to perform satisfactory regressions. An advantage of this scheme would be that medium-scale deviations from otherwise smooth trends could be pursued individually in an attempt to link growth to some fluctuating environmental condition or to some intermittent or short-lived experimental treatment.

6.4.3   Tests for the 'best' model

HURD (1977) grew *Lycopersicon esculentum* (tomato) under solution-culture conditions in a controlled environment. Treatments were combined factorially from two temperatures (17 or 20°C, constant), two light intensities (20 or 80 W m^{-2} at 400–700 nm) and two carbon dioxide concentrations (0.65 or 2.75 g m^{-3}). In each of these eight treatments the logarithms of W and L_A were fitted by HUNT and PARSONS's (1974) method. Of the sixteen resulting regressions four were linear, six were quadratic and six were cubic ($P < 0.05$). Hurd argued that this selection was unsatisfactory on biological grounds and that quadratic models should be used throughout to give a uniform family of curves.

One way of deliberately avoiding such a mixed bag of growth functions would be to stiffen the requirements for their acceptance. The relationship between the probability level for acceptance of the models and the types of model chosen is shown for Hurd's sixteen sets of data in Fig. 6–3.

The result of raising the statistical requirements for the choice of polynomial is a 'steamrollering-out' of the higher-order terms, that is, as

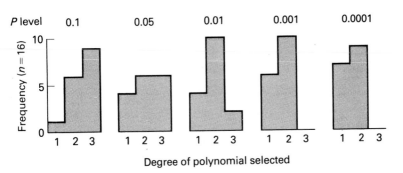

Fig. 6–3   The effect of varying the probability levels required for the acceptance of the sixteen regression models in the body of data referred to by HURD (1977).

the requirements for acceptance become progressively more stringent, cubics turn into quadratics and quadratics turn into straight lines. But at no time is a single model unquestionably the best. Hurd's suggestion that the quadratic regression might be used throughout therefore needs to be treated with caution – as a semi-mechanistic model it has its advantages but these accrue at the cost of accuracy of fit.

6.5 Extension of concepts to other fields

Although the great majority of the examples of the applications of plant growth analysis in this booklet have involved whole vascular plants, there is good reason for extending these concepts and techniques to include other systems, too. We have already seen how WILLIAMS (1975) examined the productive efficiency of RNA in young clover leaves by the calculation of a quantity analogous to unit leaf rate (section 6.2.5), and there is every possibility that other applications may be found in which the comparison of growth on an equable basis through the calculation of relative growth rates, $(1/Y)\,(dY/dX)$ in the general notation, and productive efficiencies, $(1/Z)\,(dY/dX)$, may be informative. For example, JARVIS and WILSON (1977) grew isolated embryonic axes (the plumule/radicle components of the dormant seed, with the cotyledons removed) of *Corylus avellana* (hazel) under artificial solution-culture conditions at 25°C (Fig. 6–4). Results were analysed by the methods of HUNT and PARSONS (1974). The mean axis length, from the tip of the plumule to the tip of the radicle, increased steadily in a nutrient medium alone, but in the presence also of gibberellic acid (GA_3) the early relative growth rates of the axes were stimulated dramatically (cf. the initial slopes of the two lines in Fig. 6–4).

Moving further afield, it may be that in many ways bryophytes, lichens, algae, fungi and bacteria make more accessible experimental subjects than do vascular plants since their structure is simpler and the effects of the environment on their growth and development are more direct. Again the concepts of growth of one component, in relation both to itself and to others, may be of value.

6.6 Doubling time and half life

Lower organisms often grow swiftly and values of relative growth rate, when expressed on a daily or weekly basis, tend to be high (Table 1). For this reason experimenters prefer sometimes to express such growth not as increase with time but as time taken to double in weight (or whatever): the shorter the 'doubling time' the faster the growth. A direct equivalence exists between relative growth rate and doubling time. To underline the link between this particular approach and the style of analysis adopted in the present booklet, this relationship will now be explained, using weight as the basic quantity, but on the understanding that changes in length, area, volume or number would be similarly treated, where appropriate.

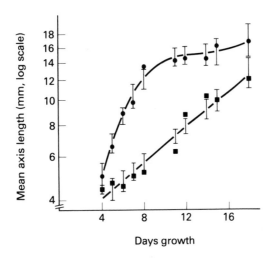

Fig. 6–4 The growth in length of excised axes of hazel seeds in a nutrient medium with (●) and without (■) gibberellic acid. (From JARVIS and WILSON, 1977.)

If the weight of an organism doubles during a given period then the difference between the logarithms of weights at the beginning and end of this period must be the logarithm of the number 2. If we use natural logarithms this difference has a value of 0.693. The question then becomes: Over what time interval do the natural logarithms of weight increase by 0.693? This time interval may be estimated graphically from a simple plot of the natural logarithms of weight against time. Alternatively, if the slope of this plot (the relative growth rate) is known, either as an instantaneous value or as a mean value over a defined interval of time, then the corresponding doubling time may be determined exactly using the relation

$$\text{Doubling time} = \frac{0.693}{R}. \qquad (6/12)$$

If the instantaneous relative growth rate, R, is used in the above equation then the instantaneous doubling time will be derived; if the mean relative growth rate, $\bar{R}$, is used then the mean time for doubling, estimated over the same time interval, will result. As we have seen, the units of R (or $\bar{R}$) are time^{-1} and so, as expected, the units of doubling time will simply be time alone.

In situations where **R** or **R̄** has a negative value, the system under investigation is not growing but decaying. Solving equation 6/12 in such cases would provide not a doubling time but a 'halving time', more commonly referred to as a half-life. This concept is familiar enough in microbiology and radiobiology and may, on occasion, be useful in plant growth analysis. For example, the decay of plant material in decomposer cycles often follows a course that may be described by a negative exponential model (MASON 1977, p. 21). This is essentially equation 3/1 running in reverse with **R** as the relative *decay* rate, and the calculation of a half-life may be a useful summary of events. Its units would again be time.

6.7 Conclusions

Time itself probably determines whether or not the concepts and techniques of plant growth analysis described in this booklet are useful to the experimenter. If he is interested in the performance of whole plants over a period of days, weeks or months, then it is almost certain that useful information can be gained by employing one or more of the methods outlined here, especially if comparisons between species or between treatments are involved. If the period of interest spans several years then these methods may become less appropriate. Real and important seasonal variation, and variation between years, may be smothered in the long-term calculation of, say, mean unit leaf rate for a temperate forest. On the other hand, the methods may be too crude to be of much use in studies dealing with performance over a matter of only minutes or hours. Section 4.6 outlined these arguments in relation to the growth of populations and communities and they may be extended along much the same lines to cover the field as a whole.

In situations where it is acknowledged that the concepts may be useful then the experimenter has the great advantage that they depend upon only the simplest of raw data acquired with the help of only the simplest of equipment. BLACKMAN (1961) pointed out that 'in the laboratory a bare minimum of a drying oven, a balance, blueprint paper, and a calculating machine will suffice' and EVANS (1972, p. 586) emphasized that 'no school is so ill-equipped that nothing could be started along these lines.' For others, working alone or in remote conditions, these advantages will be equally important; while for those working on a large scale and with the support of high-speed computers, an automated approach to plant growth analysis can become a useful routine adjunct to many experimental programmes.

6.8 Possibilities for further study

Underlining the subject's great potential, EVANS (1972, p. viii) wrote 'There are more than a quarter of a million species of higher plants, and very few of these have ever been the subject of detailed quantitative study. In few scientific fields are there such opportunites for the amateur (in the old sense of the word) to make useful contributions to the sum of ordered knowledge.' If this scope is extended, where appropriate, to include systems other than whole vascular plants then these opportunities become almost countless. The following list cites recent publications which have used the techniques of plant growth analysis to advance knowledge in a variety of fields. Some have been dealt with before in the text; some are making their first appearance here. This list provides both suggestions as to the types of study possible and examples of how each has been carried out in specific cases.

(1) Comparisons between species (GRIME and HUNT, 1975)
(2) Comparative performance of crop varieties (maize: DUNCAN and HESKETH, 1968; lettuce: SCAIFE, 1973; pine: ROBERTS and WARING, 1975)
(3) Differential effects of climate on related species (WOODWARD, 1975)
(4) Effects of CO_2 level (HUGHES and COCKSHULL, 1969)
(5) Effects of defoliation (TROUGHTON, 1973)
(6) Effects of extreme environments (Antarctic climate: SMITH and WALTON, 1975; water polluted with oil: DENNINGTON, GEORGE and WYBORN, 1975)
(7) Effects of light intensity (WATSON et al., 1972; HUGHES, 1973)
(8) Effects of mineral nutrition (BOUMA, 1970)
(9) Effects of mycorrhizal infection (STRIBLEY, READ and HUNT, 1975)
(10) Effects of seed size on subsequent growth (TAYLOR, 1972)
(11) Effects of spacing of crop plants (WATSON and FRENCH, 1971)
(12) Effects of variation in weather (WATSON, et al., 1972; THOMAS, 1975)
(13) Effects of water stress (ASHENDEN, STEWART and WILLIAMS, 1975)
(14) Growth of aquatics (MITCHELL and TUR, 1975)
(15) Growth of lichens (PROCTOR, 1977)
(16) Investigation of ontogenetic drift (EAGLES, 1969)
(17) Phenology of natural vegetation (KVĚT, SVOBODA and FIALA, 1969)

Appendix

Synopsis of symbols and formulae

Derived quantity	Contraction	Symbol	Expression for instantaneous value
Absolute growth rate	–	**G**	$\dfrac{dW}{dT}$
Biomass duration	BMD	**Z**	does not exist
Crop growth rate	CGR	**C**	$\dfrac{1}{P} \cdot \dfrac{dW}{dT}$
Leaf area duration (L_A basis)	LAD	**D**	does not exist
Leaf area duration (LAI basis)	LAD	**D**	does not exist
Leaf area index	LAI	**L**	$\dfrac{L_A}{P}$
Leaf area ratio	LAR	**F**	$\dfrac{L_A}{W}$
Leaf weight ratio	LWR	–	$\dfrac{L_W}{W}$
Rate of production, e.g. of PN per unit RNA	–	**G** $_{PN, RNA}$	$\dfrac{1}{RNA} \cdot \dfrac{dPN}{dT}$
Relative growth rate	RGR	**R**	$\dfrac{1}{W} \cdot \dfrac{dW}{dT}$
Specific absorption rate	SAR	**A**	$\dfrac{1}{R_W} \cdot \dfrac{dM}{dT}$
Specific leaf area	SLA	–	$\dfrac{L_A}{L_W}$
Specific utilization rate	SUR	**U**	$\dfrac{1}{M} \cdot \dfrac{dW}{dT}$
Unit leaf rate (=net assimilation rate)	ULR (=NAR)	**E** $_A$	$\dfrac{1}{L_A} \cdot \dfrac{dW}{dT}$
Unit shoot rate	USR	**B**	$\dfrac{1}{S_W} \cdot \dfrac{dW}{dT}$

The symbols used in the text, and in the above table, for the various measured quantities used in plant growth analysis are:
FW, total fresh weight; L_A, total leaf area; L_W, total leaf dry weight; M, mineral nutrient

Formula for mean value over the time interval $_1T$ to $_2T$ (approximate except in the case of absolute and relative growth rates)	Units	See section
$_{1-2}\bar{G} = \dfrac{_2W - _1W}{_2T - _1T}$	weight time^{-1}	3.2
$_{1-2}Z = \dfrac{(_1W + _2W)(_2T - _1T)}{2}$	weight . time	4.5.3
$_{1-2}\bar{C} = \dfrac{1}{P} \cdot \dfrac{_2W - _1W}{_2T - _1T}$	weight area^{-1} time^{-1}	4.4
$_{1-2}D = \dfrac{(_1L_A + _2L_A)(_2T - _1T)}{2}$	area . time	4.5.2
$_{1-2}D = \dfrac{(_1L + _2L)(_2T - _1T)}{2}$	time	4.5.1
$_{1-2}\bar{L} = \dfrac{_2L_A - _1L_A}{P}$	dimensionless	4.3
$_{1-2}\bar{F} = \dfrac{(_1L_A/_1W) + (_2L_A/_2W)}{2}$	area weight^{-1}	3.3
$_{1-2}\overline{LWR} = \dfrac{(_1L_W/_1W) + (_2L_W/_2W)}{2}$	dimensionless	3.4
$_{1-2}\bar{G}_{PN,\,RNA} = \dfrac{_2PN - _1PN}{_2T - _1T} \cdot \dfrac{\log_{e2}RNA - \log_{e1}RNA}{_2RNA - _1RNA}$	weight weight^{-1} time^{-1}	6.2.5
$_{1-2}\bar{R} = \dfrac{\log_{e2}W - \log_{e1}W}{_2T - _1T}$	[weight weight^{-1}] time^{-1}	3.2
$_{1-2}\bar{A} = \dfrac{_2M - _1M}{_2T - _1T} \cdot \dfrac{\log_{e2}R_W - \log_{e1}R_W}{_2R_W - _1R_W}$	weight weight^{-1} time^{-1}	6.2.3
$_{1-2}\overline{SLA} = \dfrac{(_1L_A/_1L_W) + (_2L_A/_2L_W)}{2}$	area weight^{-1}	3.4
$_{1-2}\bar{U} = \dfrac{_2W - _1W}{_2T - _1T} \cdot \dfrac{\log_{e2}M - \log_{e1}M}{_2M - _1M}$	weight weight^{-1} time^{-1}	6.2.4
$_{1-2}\bar{E} = \dfrac{_2W - _1W}{_2T - _1T} \cdot \dfrac{\log_{e2}L_A - \log_{e1}L_A}{_2L_A - _1L_A}$	weight area^{-1} time^{-1}	3.3
$_{1-2}\bar{B} = \dfrac{_2W - _1W}{_2T - _1T} \cdot \dfrac{\log_{e2}S_W - \log_{e1}S_W}{_2S_W - _1S_W}$	[weight weight^{-1}] time^{-1}	6.2.2

content (of one particular element, or of a group of elements); P, ground area; PN, protein nitrogen content; RNA, ribonucleic acid content; R_W, total root dry weight; S_W, total shoot dry weight (the above-ground parts of the plant); T, time; W, total dry weight.

Further Reading

BLACKMAN, G. E. (1961). Responses to environmental factors by plants in the vegetative phase. In *Growth in Living Systems*, ed. M. X. Zarrow. Basic Books, New York, 525–56.
   An historical account of plant growth analysis followed by a review of environmental effects.

BLACKMAN, G. E. (1968). The application of the concepts of growth analysis to the assessment of productivity. In *Functioning of Terrestrial Ecosystems at the Primary Production Level*, ed. F. E. Eckardt. Proc. Copenhagen Symp., UNESCO, 243–59.
   A production-oriented review of results obtained at the population level of organization.

CAUSTON, D. R. (1977). Plant growth analysis: a biological application of calculus. In *A Biologist's Mathematics*. Edward Arnold, London, 205–17.
   A mathematical introduction to the growth analysis of individuals. Brief but clear.

EVANS, G. C. (1972). *The Quantitative Analysis of Plant Growth*. Blackwell Scientific Publications, Oxford, pp. xxvi + 734.
   The definitive work. Advanced students of plant growth analysis will find that this covers many of the topics in the present booklet in greater depth.

KVĚT, J., ONDOK, J. P., NEČAS, J. and JARVIS, P. G. (1971). Methods of growth analysis. In *Plant Photosynthetic Production*, eds. Z. Šesták, J. Čatský and P. G. Jarvis. Dr. W. Junk, N.V., The Hague, 343–91.
   An excellent and wide-ranging review, covering both the classical and functional approaches to plant growth analysis.

RADFORD, P. J. (1967). Growth analysis formulae – their use and abuse. *Crop Sci.*, **7**, 171–5.
   A short, clear explanation of the assumptions involved in the use of the classical formulae and a brief introduction to the functional approach. Much cited by workers in agriculture.

RICHARDS, F. J. (1969). The quantitative analysis of growth. In *Plant Physiology, a Treatise, Vol. VA*, ed. F. C. Steward. Academic Press, London, 1–76.
   Provides in-depth coverage of the process of dry weight increase, from a mathematical point of view.

WATSON, D. J. (1952). The physiological basis of variation in yield. *Adv. Agron.*, **4**, 101–45.
   An extensive review of unit leaf rate ('net assimilation rate'), growth in leaf area, and leaf area index in crops.

WATSON, D. J. (1968). A prospect of crop physiology. *Ann. appl. Biol.*, **62**, 1–9.
   Briefly reviews the history of, and possible future for, growth studies on field crops.

WILLIAMS, R. F. (1975). The quantitative description of growth. In *The Shoot Apex and Leaf Growth*. Cambridge University Press, London, 9–26.
   An excellent introduction to dry weight increase and relative growth rate.

References

ALEXANDER, R. McN. (1971). *Size and Shape*. Studies in Biology no. 29, Edward Arnold, London.

ALLEN, S. E. (1974) ed. *Chemical Analysis of Ecological Materials*. Blackwell Scientific Publications, Oxford.

ASHENDEN, T. W., STEWART, W. S. and WILLIAMS, W. (1975). *J. Ecol.*, **63**, 97.

BAILEY, N. T. J. (1964). *Statistical Methods in Biology*, 2nd impression (with corrections). English Universities Press, London.

BAZZAZ, F. A. and HARPER, J. L. (1977). *New Phytol.*, **78**, 193.

BLACK, J. N. (1963). *Aust. J. agric. Res.*, **14**, 20.

BLACKMAN, G. E. and WILSON, G. L. (1951). *Ann. Bot.*, **15**, 373.

BLACKMAN, V. H. (1919). *Ann. Bot.*, **23**, 353.

BOUMA, D. (1970). *Ann. Bot.*, **34**, 1131.

BRIGGS, G. E., KIDD, F. and WEST, C. (1920a). *Ann. appl. Biol.*, **7**, 103.

BRIGGS, G. E., KIDD, F. and WEST, C. (1920b). *Ann. appl. Biol.*, **7**, 202.

BROCK, T. D. (1967). *Science*, **158**, 1012.

BROUWER, R., JENNESKENS, P. J. and BORGGREVE, G. J. (1961). *Jaarb.I.B.S. 1961*, 29.

CAUSTON, D. R. (1969). *Biometrics*, **25**, 401.

COOMBE, D. E. (1960). *J. Ecol.*, **48**, 219.

DENNINGTON, V. N., GEORGE, J. J. and WYBORN, C. H. E. (1975). *Environ. Pollut.*, **8**, 233.

DUNCAN, W. G. and HESKETH, J. D. (1968). *Crop. Sci.*, **8**, 670.

EAGLES, C. F. (1969). *Ann. Bot.*, **33**, 937.

ELIAS, C. O. and CAUSTON, D. R. (1976). *New Phytol.*, **77**, 421.

ELLIS, E. L. and DELBRUCK, M. (1939). *J. gen. Physiol.*, **22**, 365.

EMECZ, T. I. (1962). *Ann. Bot.*, **26**, 517.

EVANS, G. C. and HUGHES, A. P. (1962). *New Phytol.*, **61**, 322.

FISHER, R. A. (1921). *Ann. appl. Biol.*, **7**, 367.

GOODALL, D. W. (1949). *Ann. Bot.*, **13**, 1.

GREGORY, F. G. (1918). *3rd Ann. Rept, Exptl Res. Stn, Cheshunt*, p. 19.

GREGORY, F. G. (1926). *Ann. Bot.*, **40**, 1.

GRIME, J. P. and HUNT, R. (1975). *J. Ecol.*, **63**, 393.

HACKETT, C. (1969). *New Phytol.*, **68**, 1023.

HARPER, J. L. and WHITE, J. (1974). *Ann. Rev. Ecol. Systemat.*, **5**, 419.

HEATH, O. V. S. (1970). *Investigation by Experiment*. Studies in Biology no. 23, Edward Arnold, London.

HEATH, O. V. S. and GREGORY, F. G. (1938). *Ann. Bot.*, **2**, 811.

HUGHES, A. P. (1965). *New Phytol.*, **64**, 399.

HUGHES, A. P. (1973). *Ann. Bot.*, **37**, 267.

HUGHES, A. P. and COCKSHULL, K. E. (1969). *Ann. Bot.*, **33**, 351.

HUGHES, A. P. and EVANS, G. C. (1962). *New Phytol.*, **61**, 154.

HUGHES, A. P. and FREEMAN, P. R. (1967). *J. appl. Ecol.*, **4**, 553.

HUNT, R. (1973). *J. appl. Ecol.*, **10**, 157.

HUNT, R. (1975). *Ann. Bot.*, **39**, 745.

HUNT, R. (1978). *New Phytol.*, **80**, 269.

HUNT, R. and BURNETT, J. A. (1973). *Ann. Bot.*, **37**, 519.

HUNT, R. and PARSONS, I. T. (1974). *J. appl. Ecol.*, **11**, 297.

HUNT, R. and PARSONS, I. T. (1977). *J. appl. Ecol.*, **14**, 965.

HURD, R. G. (1977). *Ann. Bot.*, **41**, 779.

JARVIS, B. C. and WILSON, D. (1977). *New Phytol.*, **78**, 397.

JARVIS, P. G. and JARVIS, M. S. (1964). *Physiologia Pl.*, **17**, 654.

KEAY, J., BIDDISCOMBE, E. F. and OZANNE, P. G. (1970). *Aust. J. agric. Res.*, **21**, 33.

KEYFITZ, N. (1968). *Introduction to the Mathematics of Population*. Addison-Wesley, Reading, Mass.

KREUSLER, U., PREHN, A. and HORNBERGER, R. (1879). *Landw. Jbr.*, **8**, 617.

KVĚT, J., SVOBODA, J. and FIALA, K. (1969). *Hydrobiologia*, **10**, 63.

MACHIN, D. (1976). *Biomathematics: an Introduction*. Macmillan, London.

MASON, C. F. (1977). *Decomposition*. Studies in Biology no. 74, Edward Arnold, London.

MILNER, C. and HUGHES, R. E. (1968). *Methods for the Measurement of the Primary Production of Grassland*, IBP Handbook no. 6, Blackwell Scientific Publications, Oxford.

MITCHELL, D. S. and TUR, N. M. (1975). *J. appl. Ecol.*, **12**, 213.

NELDER, J. A. (1975). *Computers in Biology*. Wykeham, London.

NICHOLLS, A. O. and CALDER, D. M. (1973). *New Phytol.*, **72**, 571.

NYE, P. H., BREWSTER, J. L. and BHAT, K. K. S. (1975). *Pl. Soil.*, **42**, 161.

ONDOK, J. P. (1971). *Photosynthetica*, **5**, 269.

OTSUKI, Y., SHIMOMORA, T. and TAKEBE, I. (1972). *Virology*, **50**, 45.

PARKER, R. F. (1978). *Introductory Statistics for Biology.* Studies in Biology no. 43, 2nd edn. Edward Arnold, London.

PEARSALL, W. H. (1927). *Ann. Bot.*, **41**, 549.

PHILLIPSON, J. (1966). *Ecological Energetics*. Studies in Biology no. 1, Edward Arnold, London.

PROCTOR, M. C. F. (1977), *New Phytol.*, **79**, 659.

RAJAN, A. K., BETTERIDGE, B. and BLACKMAN, G. E. (1973). *Ann. Bot.*, **37**, 287.

RICHARDS, F. J. (1959). *J. exp. Bot.*, **10**, 290.

ROBERTS, J. and WAREING, P. F. (1975). *Ann. Bot.*, **39**, 311.

SARUKHÁN, J. and HARPER, J. L. (1973). *J. Ecol.*, **61**, 675.

SCAIFE, M. A. (1973). *Ann. appl. Biol.*, **74**, 119.

ŠESTÁK, Z., ČATSKÝ, J. and JARVIS, P. G. (1971). eds. *Plant Photosynthetic Production*, Dr. W. Junk N.V., The Hague.

SINGH, J. S., LAUENROTH, W. K. and STEINHORST, R. K. (1975). *Bot. Rev.*, **41**, 181.

SMITH, R. I. L. and WALTON, D. W. H. (1975). *Ann. Bot.*, **39**, 831.

SOLOMON, M. E. (1976). *Population Dynamics*. Studies in Biology no. 18, 2nd edn. Edward Arnold, London.

STOY, V. (1965). *Physiologia Pl. Suppl. IV.*, 1.

STRIBLEY, D. P., READ, D. J. and HUNT, R. (1975). *New Phytol.*, **75**, 119.

TAYLOR, G. B. (1972). *Aust. J. agric. Res.*, **23**, 595.

THOMAS, H. (1975). *J. agric. Sci., Camb.*, **84**, 333.

THORNE, G. N. (1961). *Ann. Bot.*, **25**, 29.

THORNLEY, J. H. M. (1976). *Mathematical Models in Plant Physiology*. Academic Press, London.

TRINCI, A. P. J. (1969). *J. gen. Microbiol.*, **57**, 11.

TROUGHTON, A. (1955). *Agric. Prog.*, **30**, 59.

TROUGHTON, A. (1956). *J. Br. Grassld Soc.*, **11**, 56.

TROUGHTON, A. (1973). *Pl. Soil.*, **38**, 95.

VERNON, A. M. and ALLISON, J. C. S. (1963). *Nature, Lond.*, **200**, 814.

WATSON, D. J. (1947). *Ann. Bot.*, **11**, 41.

WATSON, D. J. (1958). *Ann. Bot.*, **22**, 37.

WATSON, D. J. (1971). In *Potential Crop Production*, eds. P. F. Wareing and J. P. Cooper. Heinemann, London, p. 76.

WATSON, D. J. and FRENCH, S. A. W. (1971). *J. appl. Ecol.*, **8**, 421.

WATSON, D. J., MOTOMATSU, T., LOACH, K. and MILFORD, G. F. J. (1972). *Ann. appl. Biol.*, **71**, 159.

WELBANK, P. J. (1962). *Ann. Bot.*, **26**, 361.

WELBANK, P. J. (1964). *Ann. Bot.*, **28**, 1.

WEST, C., BRIGGS, G. E. and KIDD, F. (1920). *New Phytol.*, **19**, 200.

WHALEY, W. G. (1961). In *Encyclopedia of Plant Physiology XIV*, ed. W. Ruhland, Springer-Verlag, Berlin, p. 71.

WHITEHEAD, F. H. and MYERSCOUGH, P. J. (1962). *New Phytol.*, **61**, 314.

WILLIAMS, R. F. (1946). *Ann. Bot.*, **10**, 41.

WILLIAMS, R. F. (1948). *Aust. J. sci. Res. Ser. B.*, **1**, 333.

WOODWARD, F. I. (1975). *New Phytol.*, **74**, 335.